HILBERT'S THIRD PROBLEM

D1480264

SCRIPTA SERIES IN MATHEMATICS

516.23
B639h

HILBERT'S THIRD PROBLEM

Vladimir G. Boltiânskiĭ
Steklov Institute of Mathematics,
USSR Academy of Sciences

Translated by
Richard A. Silverman

and introduced by
Albert B. J. Novikoff
New York University

1978
V. H. WINSTON & SONS
Washington, D.C.

A HALSTED PRESS BOOK

JOHN WILEY & SONS
New York   Toronto   London   Sydney

Copyright © 1978, by V. H. Winston & Sons, a Division of
Scripta Technica, Inc.

All rights reserved. Printed in the United States of America.
No part of this publication may be reproduced, stored in a
retrieval system, or transmitted, in any form or by any means,
electronic, mechanical, photocopying, recording, or otherwise,
without prior written permission of the publisher.

V. H. Winston & Sons, a Division of Scripta Technica, Inc.,
Publishers
1511 K St. N.W., Washington, D.C. 20005

Distributed solely by Halsted Press, a Division of John Wiley
& Sons, Inc.

Library of Congress Cataloging in Publication Data
Boltiānskiĭ, Vladimir Grigor'evich.
   Hilbert's third problem.

   (Scripta series in mathematics)
   Translation of Tret'iā problema Gil'berta.
   Bibliography: p.
   1. Tetrahedra.   2. Hilbert, David, 1862-1943.
I. Title.   II. Series.
QA491.B6213   516'.23   77-19011
ISBN 0-470-26289-3

Composition by Isabelle Sneeringer, Scripta Technica, Inc.

CONTENTS

v

ALLEGHENY COLLEGE LIBRARY

79-1942

PREFACE TO
THE AMERICAN EDITION

The American reader is already familiar with translations of several of my books, and it is a pleasure to know that the present volume, an effort very dear to me, has been translated into English, as well.

Mathematics is my hobby, as well as my profession. I am especially fond of geometry and all my work, even papers on control theory, embody the geometric point of view, which explains why this book was written. I first came across the equidecomposability theory in a small booklet by Prof. V. F. Kagan, given to me as a prize at the Sixth Moscow High School Mathematical Olympics in 1940. Later, under the influence of the remarkable geometric studies of Prof. H. Hadwiger and his colleagues, I have written (in 1956) a popular exposition of the subject in the small book *Equivalent and Equidecomposable Figures.** Since that time, however, much new work has been done on Hilbert's third problem, and I began carrying in the back of my mind the idea that a systematic, up-to-date treatment of this subject was perhaps in order.

Then, not long ago, A. V. Pogorelov's *Hilbert's Fourth Problem*** appeared in Moscow. Its title gave me the stimulus needed for writing the present book. I worked with enthusiasm. On January 26, 1976, I put a ream of clean paper on my table at the ice-skating resort of

*D. C. Heath, 1963.
**To appear soon in English translation in the Scripta Series in Mathematics.

Ramon' near Voronezh. Four months later, the book was finished. Now its translation is before American readers. I hope that it will serve as one small brick in the edifice of cultural cooperation and friendly relationships between our peoples.

I am very grateful to the translator-editor of the book, Dr. Richard Silverman, the Series Editor, Prof. Irwin Kra, and to all those who contributed to its publication. My thanks also go to V. H. Winston & Sons for initiating the project, and to John Wiley & Sons, with whom I have cooperated in the past.

V. G. Boltiãnskiĭ

FOREWORD

The list of problems accessible to undergraduate students that lead in a natural way to significant, even thrilling mathematical developments is regrettably small. Number theory seems to be particularly rich in such problems, of which Fermat's last theorem is the outstanding example. Euler's Koenigsberg Bridge problem is another from an entirely different source. To cite further examples, this time from geometry, the definition of congruence via transformation groups leads immediately to Klein's circle of ideas, while the ancient question of the independence of the parallel postulate leads ultimately to elliptic and hyperbolic geometries.

The study of "planimetry," the measurement of areas of polygonal figures, leads easily to the question of determining when two polygons are "equidecomposable," that is, when can they both be decomposed into the same (finite) number of pairwise congruent polygonal "pieces"? Is the obvious necessary condition that they have the same area also sufficient? In other words, given two polygons of equal area, are they both solutions of a common jigsaw puzzle? The answer happens to be yes, as proved by the elder Bolyai (see Theorem 10 of this book). The problem is not a perverse exercise in ingenuity; on the answer depends whether or not there exists an "elementary" theory of planimetry, which exploits only the area formula for rectangles, but makes no appeal to limit processes such as Archimedean "exhaustion" or Cavalieri's principle.

One is then led to formulate, as did Hilbert, the analogous question for polyhedra, which is in fact the third of his celebrated list of 23 problems presented in 1900. In this three-dimensional case the answer turns out to be no, as anticipated by Hilbert from the observation that no known proof of the formula for the volume of a pyramid was free of limit processes. The negative answer was first proved by Max Dehn (1900), following a clue discovered by Bricard. Dehn's method of proof has been vastly simplified and generalized, largely through the efforts of the Swiss school of geometers, and can now be put in an unexpected and modern algebraic form. Hilbert's Third Problem has thus led to a fascinating interplay between special cases and general theories, each in the service of the other.

It is remarkable that this elegant and fruitful evolution, from such an innocent-seeming problem, remains unknown except to a restricted circle of specialists, few of whom publish in English. Prof. Boltiãnskiĭ has placed the mathematical community in his debt by providing an eminently readable and completely self-contained account of all this, which his translator Dr. Richard Silverman has rendered in equally accessible English. Moreover, the book is exceptionally browsable. All proofs may be skipped on first reading, as the assertions of the individual theorems are complete in themselves. The many digressions, orienting remarks, historical discussions and the like are set apart from the formal proofs, inviting the reader to get the main gist before attacking the details. The result is an outstanding example of the confluence of elementary (but deep) mathematics, brilliant pedagogy and modern research.

Prof. Albert B. J. Novikoff

INTRODUCTION

As the nineteenth century gave way to the twentieth, the Second International Congress of Mathematicians was held in Paris. At one session of the congress (Aug. 8, 1900), David Hilbert read his famous report entitled "Mathematical Problems" [33]. This remarkable work, written by a prominent mathematician with the most varied interests, comprises all the main directions of mathematics. Interest in Hilbert's problems has remained strong throughout the twentieth century, and it would be hard to overestimate the significance of the problems for the development of mathematics.

The great majority of the 23 problems posed by Hilbert pertain to new, rapidly developing branches of mathematics (to many of which Hilbert himself made important contributions). And only one problem, the third, deals with questions related to the teaching of "high school" geometry. Hilbert calls attention to the fact that as far back as the time of Euclid, the volume of a triangular pyramid has been calculated by using a rather complicated limiting process (the "devil's staircase"; cf. Fig. 1, taken from the geometry text by A. P. Kiselev [40], p. 53), whereas the study of areas in planimetry manages to avoid an analogous limiting process. The essence of the problem is to justify the use of this "superfluous" (as compared with planimetry) limiting processes, and to *prove* that, without the use of such a process, one cannot construct a theory of volume for polyhedra.

1

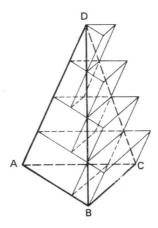

FIG. 1

It was not only because of his interest in pedagogical questions of elementary geometry that Hilbert posed this problem. (It will be recalled that the report on "Mathematical Problems" was made in the year following the publication of the book "Foundations of Geometry" [32], in which Hilbert presented his widely known system of axioms for Euclidean geometry.) Hilbert obviously foresaw that the significance of the problem he had posed far transcended the narrow context of pedagogical methodology in matters involving volumes and that the problem might in fact lead to the creation of a theory of equidecomposability of polyhedra, which is both mathematically interesting and rich in results.

Hilbert's foresight has been completely justified. In the very same year 1900, M. Dehn [12] confirmed Hilbert's conjecture by proving the existence of polyhedra of equal volume which are not equidecomposable. Dehn's very complicated (and, it must be said, very confusing) proofs were substantially improved by B. F. Kagan [39] back in 1903. But the middle of the twentieth century has also been distinguished by new results along these lines. In fact, the famous Swiss geometer H. Hadwiger, together with his students, has introduced fresh new ideas into the theory of equidecomposability (see [20]–[27], [30] and a number of later papers).

The author's book [6] contains the simplest proof of Dehn's theorem, based on the ideas of the Swiss school of geometers, as well as an exposition of the most important results connected with Hilbert's third problem. At the time of its appearance (in 1956), this book (see also the encyclopedia article [7]) contained a relatively complete review of results relating to this direction of research. However, the book has long since gone out of print, and the last two decades have seen further progress in this field. In particular, J.-P. Sydler [54] has shown that the invariants found by Dehn give not only a necessary condition for equidecomposability, but also a sufficient condition. The investigation of questions associated with Hilbert's third problem can now be regarded as complete for the plane and three-dimensional space, and as almost complete for four-dimensional space. It is this that has compelled me to write a new book, acquainting the reader with the contemporary state of the theory.

The book is intended for mathematics teachers and pupils, as well as those actively engaged in research. The material in Sections 1-3, 5-7 and 9-14 is also accessible to high school students interested in mathematics. I have tried to introduce the reader to the contemporary state of knowledge in the subject matter area of the book, while at the same time keeping the level of exposition as elementary as possible.

I would like to take this occasion to express my deep gratitude to N. V. Yefimov, V. A. Zalgaller, and I. M. Yaglom for their friendly concern and valuable advice.

V. G. Boltiãnskiĭ

CHAPTER 1

THE MEASUREMENT
OF AREA AND VOLUME[1]

§1. The concept of area

By the area $s(F)$ of a figure F one ordinarily means the number of units of area "making up" the figure F, where the unit of area is taken to be the square whose side is the segment chosen as the unit of length. However, this interpretation (accepted in high school) can hardly serve as a precise mathematical *definition* of the concept of area, but only as an intuitive explanation, suggestive of ideas drawn from everyday life. For example, it is not clear how one can define the number of units of area "making up" a circular disk of given radius.

One way of making the concept of area more precise starts by considering a *mosaic*, i.e., a decomposition of the plane into congruent squares. Suppose, for example, that the side of each of the

---

[1] An exposition of the theory of area and volume can be found in many sources; see, for example, references [28] and [46], as well as I. M. Yaglom's supplement to the interestingly written little book by Dubnov [13]. If he wishes, the reader can follow the line of thought in Chapter I only cursorily, paying somewhat greater attention to the last three sections of the chapter.

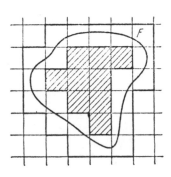

FIG. 2

squares forming the mosaic is of length 1. In Fig. 2 the figure F contains a figure made up of 9 squares of the mosaic, and is itself contained in a figure made up of 29 squares. This allows us to write the double inequality

$$9 \leqslant s\,(F) \leqslant 29.$$

For a more precise estimate of the area we can use a mosaic whose squares have sides of length 1/10 (so that each square of the previous mosaic contains 100 squares of the new mosaic). If, say, F contains a figure made up of 1716 squares of the new mosaic, and is contained in a figure made up of 1925 such squares, then

$$17.16 \leqslant s\,(F) \leqslant 19.25.$$

Refining the mosaic once again (i.e., decreasing the lengths of the sides of the squares by a factor of 10), we get an even more accurate estimate of the area of the figure F, and so on.

To make the process just described more precise, we agree to regard the initial mosaic (whose squares have sides of length 1) as the *zeroth mosaic*. After k refinements we get the kth mosaic, whose squares have sides of length $1/10^k$. Suppose F contains a figure made up of a_k squares of the kth mosaic, and is contained in a figure made up of b_k such squares, where $k = 0, 1, 2, \ldots$. Since every square of the kth mosaic contains 100 squares of the $(k + 1)$st mosaic, after refinement the a_k squares of the kth mosaic give $100a_k$ squares of the $(k + 1)$st mosaic contained in F. Moreover, F can contain some extra squares of the $(k + 1)$st mosaic (obtained by refinement of the squares of the kth mosaic which were not entirely contained in F). Therefore $a_{k+1} \geqslant 100a_k$, and similarly $b_{k+1} \leqslant 100b_k$. Dividing these inequalities by $10^{2(k+1)}$, we obtain

$$\frac{a_k}{10^{2k}} \leqslant \frac{a_{k+1}}{10^{2(k+1)}} , \qquad \frac{b_{k+1}}{10^{2(k+1)}} \leqslant \frac{b_k}{10^{2k}} , \qquad k = 0, 1, 2, \ldots .$$

Thus

$$a_0 \leqslant \frac{a_1}{10^2} \leqslant \frac{a_2}{10^4} \leqslant \cdots \leqslant \frac{a_k}{10^{2k}} \leqslant \cdots \leqslant \frac{b_k}{10^{2k}} \leqslant \cdots$$

$$\cdots \leqslant \frac{b_2}{10^4} \leqslant \frac{b_1}{10^2} \leqslant b_0, \qquad (1)$$

so that the limits

$$\lim_{k \to \infty} \frac{a_k}{10^{2k}} = \underline{s} \, (F), \qquad \lim_{k \to \infty} \frac{b_k}{10^{2k}} = \bar{s} \, (F), \qquad (2)$$

exist, with $\underline{s} \, (F) \leqslant \bar{s} \, (F)$. If the figure F is such that these limits coincide, then we call the figure F *measurable* (or "squarable"), and the number $\underline{s} \, (F) = \bar{s} \, (F)$, i.e., the common value of the limits (2), is called the *area*[2] of the figure F, denoted by $s \, (F)$. In other words, the process of measurement just described is not only used to *estimate* the area (or to *calculate* it with the help of a limiting process), but also serves as the very *definition* of the concept of area.

There is, however, one difficulty with the above definition. Once the zeroth mosaic has been chosen, the first mosaic is uniquely defined. In fact, the sides of every square of zeroth mosaic are

---

[2] This concept of area is associated with the work of Jordan, and hence $s \, (F)$ is sometimes called *Jordan content* or *Jordan area* (see [28], Chap. III).

divided into 10 congruent parts, and then lines are drawn through the points of subdivision parallel to the lines of the zeroth mosaic. In just the same way, the first mosaic uniquely determines the second mosaic (by division of the sides of its squares into 10 parts), and so on. Thus, if the zeroth mosaic is *fixed*, all the subsequent mosaics are uniquely determined, and hence the numbers a_k, b_k and the limits (2) are uniquely determined (for a given figure F). However, if the zeroth mosaic is shifted or rotated, then the numbers a_k and b_k can change. Thus in Fig. 3 the figure φ contains 5 squares of the initial mosaic (indicated by the solid lines) and only 3 squares of the displaced mosaic (indicated by the dashed lines). It is not clear a *a priori* that the concept of measurability is invariant under displacement of the zeroth mosaic (i.e., perhaps a figure measurable with one position of the zeroth mosaic can turn out to be nonmeasurable with another position). It is also not clear *a priori* that the area of a figure F calculated with one position of the zeroth mosaic will coincide with the area of the same figure calculated with another position of the zeroth mosaic (provided that F is measurable in both cases).

This difficulty can be overcome by rigidly fixing all the mosaics. For example, fixing a rectangular system of coordinates x and y, we can regard the kth mosaic as the system of squares into which the plane is partitioned by the lines $x = p/10^k$, $y = q/10^k$, where p and q range over the integers. For the time being, we will confine

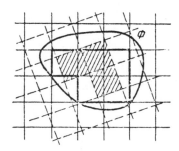

FIG. 3

ourselves to this concept of area. Later on, we will talk about what happens when the zeroth mosaic is shifted or rotated.

We note that the area is a *function s*, defined on the set of all measurable figures: To every figure *F* belonging to this set (that is, to every measurable *F*) the function assigns a real number *s(F)*.

§2. The area axioms

The definition presented in the preceding section allows us to prove a number of *properties* of area. In this section, we consider only the following four primary properties, which it will subsequently be convenient to regard as *axioms*:

(α) *The function s is nonnegative, i.e., the area s(F) of any measurable figure F is a nonnegative number.*

(β) *The function s is additive, i.e., if F' and F'' are measurable figures, with no common interior points, then the figure F' $\cup$ F'' is also measurable and s (F' $\cup$ F'') = s (F') + s (F'').*

(γ) *The function s is invariant under translations (parallel displacements), i.e., if F is a measurable figure and F' is the figure obtained by subjecting F to a translation, then the figure F' is also measurable and s (F') = s (F).*

(δ) *The function s is normalized, i.e., the unit square Q is a measurable figure and s(Q) = 1.*

By the unit square in property (δ) we mean a fixed square whose sides are of length 1, namely one of the squares of the zeroth mosaic.

In property (β) we talk about figures which have no common interior points. In this regard, let us adopt the following notation: The union of k figures $F_1, \ldots, F_k$, no two of which have common interior points, will be denoted by $F_1 + \ldots + F_k$. In other words, the notation $F = F_1 + \ldots + F_k$ means that the following two conditions are satisfied: 1) The interiors of the figures $F_1, \ldots, F_k$ are pairwise disjoint; 2) $F = F_1 \cup \ldots \cup F_k$. With this notation we can express property (β) as follows: If $F = F' + F''$, then $s (F) = s (F') + s (F'')$. From this property we can deduce the following more general proposition, with the help of an obvious induction: If $F = F_1 + \ldots + F_k$, then $s(F) = s(F_1) + \ldots + s(F_k)$.

Proof of property (α). The number a_0 (i.e., the number of squares of the zeroth mosaic contained in the figure F) is either zero or a positive integer. It follows from the inequality $a_0 \geqslant 0$ that the numbers $a_k/10^{2k}$ are all nonnegative (see (1)) and therefore $\underline{s}(F) \geqslant 0$, by (2). Since the figure F is measurable, we have $s(F) = \underline{s}(F) \geqslant 0$.

Proof of property (β). The numbers a_k and b_k, constructed for the figure F', will now be denoted by a'_k and b'_k. In other words, F' contains a figure made up of a'_k squares of the kth mosaic, and is contained in a figure made up of b'_k such squares. We denote the analogous numbers for the figure F'' by a''_k and b''_k, reserving the notation a_k, b_k (without primes) for the figure $F' + F''$.

No square can be simultaneously contained in both the figure F' and the figure F''. Therefore, taking all the squares of the kth mosaic contained in F', and all the squares of the kth mosaic contained in F'', we get a figure made up of $a'_k + a''_k$ squares of the mosaic and contained in $F' + F''$. There may also be squares of the kth mosaic which are contained in neither of the figures F' and F'', but which are contained in $F' + F''$ (Fig. 4). Thus $a_k \geqslant a'_k + a''_k$, and hence

$$\frac{a_k}{10^{2k}} \geqslant \frac{a'_k}{10^{2k}} + \frac{a''_k}{10^{2k}}.$$

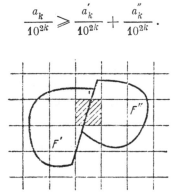

FIG. 4

This inequality is valid for every $k = 0, 1, 2, \ldots$. Taking the limit as $k \to \infty$ (see (?)), we obtain

$$\underline{s}\ (F' + F'') \geqslant s\ (F') + s\ (F'').$$

Similarly, considering the numbers b'_k, b''_k, b_k, we find that

$$\overline{s}\ (F' + F'') \leqslant s\ (F') + s\ (F'').$$

Since $\underline{s}\ (F' + F'') \leqslant \overline{s}\ (F' + F'')$ as always, we have

$$\underline{s}\ (F' + F'') = \overline{s}\ (F' + F'') =$$
$$= s\ (F') + s\ (F'').$$

But this means that property (β) holds.

Proof of property (γ). Let Q' be the square obtained by subjecting the unit square Q to a translation, and let q_0 be the vertex of the *first* mosaic which is contained in the square Q' and is closest to the lower left-hand vertex of Q' (Fig. 5). The horizontal line going through q_0 is at a distance of less than $1/10$ from the lower side of the square Q', and the same is true of the vertical line through q_0 and the left side of Q'. It follows that the square of side length $9/10$, whose lower left-hand vertex is the point q_0, is contained in Q' (see Fig. 5), i.e., Q' contains a figure made up of 81 squares of the first

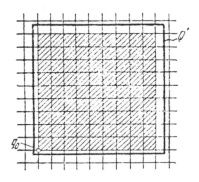

FIG. 5

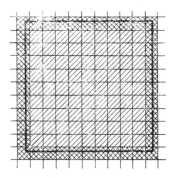

FIG. 6

mosaic. (Note that Q' may contain a larger number of squares of the first mosaic; this happens if the point q_0 lies on the boundary of the square Q'.) Bordering the square of side 9/10 contained in Q' with a strip of width equal to one square of the first mosaic (Fig. 6), we obtain a figure containing Q' and made up of 121 squares of the first mosaic. A similar argument, applied to the kth mosaic, shows that Q' contains a figure made up of $(10^k - 1)^2$ squares of the kth mosaic, and is contained in a figure made up of $(10^k + 1)^2$ such squares. In other words, the numbers a_k and b_k (chosen for the square Q') satisfy the inequalities $a_k \geqslant (10^k - 1)^2$, $b_k \leqslant (10^k + 1)^2$, and therefore

$$\frac{a_k}{10^{2k}} \geqslant \left(1 - \frac{1}{10^k}\right)^2, \quad \frac{b_k}{10^{2k}} \leqslant \left(1 + \frac{1}{10^k}\right)^2.$$

Taking the limit as $k \to \infty$ (see (2)), we get $\underline{s}\,(Q') \geqslant 1$, $\bar{s}\,(Q') \leqslant 1$, i.e., $\underline{s}\,(Q') = \bar{s}\,(Q') = 1$. Thus Q' is a measurable figure and $s\,(Q') = 1$.

Similarly, if P is a square of the kth mosaic, and if P' is the square obtained by subjecting P to a translation, then P' is a measurable figure and $s\,(P') = 1/10^{2k}$.

Next, let G be a figure made up of a squares of the kth mosaic, and let $G' = t(G)$ be the figure obtained from G by subjecting G to the translation t. Let $P_1, \ldots, P_a$ be the squares of the k-th mosaic making up G, so that $G = P_1 + \ldots + P_a$. By the property (β) proved earlier, we have $s(G) = s(P_1) + \ldots + s(P_a) = a/10^{2k}$. Since every square $t(P_i)$ is of area $1/10^{2k}$, as shown above, we have the analogous equality $s(G') = s(t(G)) = s(t(P_1)) + \ldots + s(t(P_a)) = a/10^{2k}$. Thus $s(G) = s(G')$, i.e., the area of a figure consisting of a set of squares from the kth mosaic is invariant under translation.

Finally, let F be an arbitrary measurable figure. Since $\underline{s}(F) = s(F)$, then, given any $\varepsilon > 0$, we can find a k such that $a_k/10^{2k} > s(F) - \varepsilon/2$, where a_k is such that some figure G made up of a_k squares of the kth mosaic is contained in F. In other words, $s(G) > s(F) - \varepsilon/2$, where $G \subset F$ and the figure G consists of squares from the kth mosaic. We now apply a translation t. Then $t(G) \subset t(F)$, where, as just proved, the figure $t(G)$ is measurable and $s(t(G)) = s(G)$. The fact that the figure $t(G)$ is measurable implies the existence of an l such that there is a figure G^*, made up of squares of the lth mosaic, which satisfies the conditions $G^* \subset t(G)$ and $\overset{\bullet}{a_l}/10^{2l} > s(t(G)) - \varepsilon/2$, where $\overset{\bullet}{a_l}$ is the number of squares of the lth mosaic making up the figure G^*. We have

$$G^* \subset t(G) \subset t(F),$$

$$\frac{\overset{\bullet}{a_l}}{10^{2l}} > s(t(G)) - \frac{\varepsilon}{2} = s(G) - \frac{\varepsilon}{2} > s(F) - \varepsilon.$$

Thus the figure G^*, made up of $\overset{\bullet}{a_l}$ squares of the lth mosaic, is contained in $t(F)$, and hence $\underline{s}(t(F)) \geqslant \overset{\bullet}{a_l}/10^{2l} > s(F) - \varepsilon$. Since this holds for arbitrary $\varepsilon > 0$, we have $\underline{s}(t(F)) \geqslant s(F)$.

An analogous argument establishes the inequality $\bar{s}(t(F)) \leqslant s(F)$. But $\underline{s}(t(F)) \leqslant \bar{s}(t(F))$, as always, and hence $\underline{s}(t(F)) = \bar{s}(t(F)) = \bar{s}(t(F)) = s(F)$. This completes the proof of property (γ).

The proof of property (δ) is obvious.

Before turning to a discussion of the axiomatic approach to the concept of area, we prove that every *polygon* is measurable. By a polygon we mean a bounded closed set in the plane, whose boundary is the union of a finite number of line segments. It follows from this definition that figures bounded by one or several closed polygonal lines are regarded as polygons, including *disconnected* figures consisting of separate pieces (Fig. 7). A polygon can also be defined as the union of a finite number of triangles (this definition is equivalent to the one just given).

Theorem 1. *Every polygon is a measurable figure.*

PROOF. Let L denote the polygonal line (possibly consisting of several separate pieces) forming the boundary of the polygon F, and let p be the length of L. We fix some positive integer k and construct a system of points $x_1, \ldots, x_q$ on the line L making up a $1/10^k$-net (i.e., such that every point $x \in L$ is at a distance less than $1/10^k$ from one of the points x_i). It is not hard to see that the number q can be regarded as satisfying the condition $q \leqslant p \cdot 10^k + m$, where m is the number of segments ("links") of which L consists. In fact, let $L_1, \ldots, L_m$ be the links of the polygonal line L, and let p_i be the length of the link L_i, so that $p = p_1 + \ldots + p_m$. By consecutively laying off segments of length $1/10^k$ along the segment L_i (starting from one of its ends), we get a $1/10^k$-net on L_i consisting of no more than $p_i \cdot 10^k + 1$ points. Choosing such $1/10^k$-nets on each of the links $L_1, \ldots, L_m$, we get a $1/10^k$-net on the line L, consisting of no more than $(p_1 + \ldots + p_m)10^k + m = p \cdot 10^k + m$ points.

Thus suppose that a $1/10^k$-net $\{x_1, \ldots, x_q\}$ is chosen on L, with $q \leqslant p \cdot 10^k + m$, and let Q_i denote the union of the nine squares of the kth mosaic, one of which contains the point x_i and is bordered by the other eight (Fig. 8). All the points of the line L at a

FIG. 7

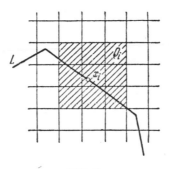

FIG. 8

distance less than $1/10^k$ from x_i are contained inside the square Q_i. Therefore all the squares of the kth mosaic having points in common with the line L are contained in the figure $Q_1 \cup \ldots \cup Q_q$, which is made up of no more than $9q$ squares of the kth mosaic. Hence the number of squares of the kth mosaic having points in common with the line L does not exceed $9q \leqslant 9 \, (p \cdot 10^k + m)$.

Now let a_k be the number of squares of the kth mosaic contained in the polygon F, and let b_k be the number of squares of the kth mosaic having points in common with F. It is clear that every square having points in common with F but not included among the a_k squares entirely contained in F must have points in common with the line L, so that

$$0 \leqslant b_k - a_k \leqslant 9 \, (p \cdot 10^k + m).$$

Thus

$$0 \leqslant b_k/10^{2k} - a_k/10^{2k} \leqslant 9p/10^k + 9m/10^{2k}$$

which gives $0 \leqslant \bar{s}\,(F) - \underline{s}\,(F) \leqslant 0$ after taking the limit as $k \to \infty$. But then $\underline{s}\,(F) = \bar{s}\,(F)$, i.e., the polygon F is a measurable figure.

ALLEGHENY COLLEGE LIBRARY

Theorem 2. *There exists one and only one function s defined on the set of all polygons which satisfies the conditions* (α), (β), (γ), (δ).

PROOF. Since every polygon F is a measurable figure (Theorem 1), its area $s(F)$ is defined, and, as we have seen, this function satisfies the conditions (α), (β), (γ), (δ). This proves the existence of the function s.

To prove the uniqueness of s, let s^* be any other function which is defined on all polygons and satisfies the conditions (α), (β), (γ), (δ). According to (δ), $s^*(Q) = 1$, where Q is the unit square. We now consider the 100 squares of the first mosaic contained in Q, and denote them by $P_1, \ldots, P_{100}$. Given any two of these squares, each can be obtained from the other with the help of a translation, and therefore, by (γ), $s^*(P_1) = s^*(P_2) = \ldots = s^*(P_{100})$. Moreover, since $Q = P_1 + \ldots + P_{100}$, it follows from (β) that $s^*(P_1) + \ldots + s^*(P_{100}) = s^*(Q) = 1$, i.e., $100s^*(P_1) = 1$, and hence $s^*(P_1) = 1/100$, where P is an arbitrary square of the first mosaic. Similarly, it can be shown that if P is any square of the kth mosaic, then $s^*(P) = 1/10^{2k}$. Hence, if G is a figure made up of a squares of the kth mosaic, we have $s^*(G) = a/10^{2k}$. In other words, the formula $s^*(G) = s(G)$ holds for every figure G made up of squares of the kth mosaic.

Finally, let F be an arbitrary polygon. Let G_k denote the union of all the squares of the kth mosaic contained in F, and let a_k be the number of squares of the kth mosaic making up the figure G_k. Moreover, let H_k denote the polygon such that $G_k + H_k = F$ (i.e., $H_k = \overline{F \setminus G_k}$). By (β), we have $s^*(F) = s^*(G_k) + s^*(H_k)$, which implies $s^*(F) \geqslant s^*(G_k)$ since, according to (α), $s^*(H_k) \geqslant 0$. But, as just proved, $s^*(G_k) = s(G_k) = a_k/10^{2k}$, and hence $s^*(F) \geqslant a_k/10^{2k}$. Taking the limit as $k \to \infty$, we get $s^*(F) \geqslant \underline{s}(F)$, i.e., $s^*(F) \geqslant s(F)$. Similarly, we can prove the inequality $s^*(F) \leqslant s(F)$. Therefore $s^*(F) = s(F)$ for an arbitrary polygon F.

The theorem just proved shows that for polygons the constructive definition of area, presented in the preceding section, can be replaced the following axiomatic definition: *By the area is meant a real function defined on the set of all polygons and satisfying the conditions* (α), (β), (γ), (δ) (which, with this approach, are regarded as *axioms* for area). Theorem 2 shows that this axiomatization is consistent and complete (since the function s exists and is unique). With the axiomatic definition there is no need to consider mosaics, but the unit square Q is *fixed*. The problem of changing the unit square will be considered in the next section.

The theory presented in this book, containing a solution of Hilbert's third problem and the development of the ideas implicit in the problem, is restricted to the study of the area of *polygons*[3] (and the volume of *polyhedra*, in space). However, for completeness we now show how the axiomatic definition of area can be extended to the case of arbitrary measurable figures.

Theorem 3. *A figure F is measurable if and only if given any* $\varepsilon > 0$, *there exist polygons G and H such that* $G \subset F \subset H$ *and* $s(H) - s(G) < \varepsilon$.

PROOF. Suppose the figure F is measurable. Then $\bar{s}(F) = \underline{s}(F)$ (see (2)), and hence, given any $\varepsilon > 0$, we can find a positive integer k such that there exists a figure G_k, made up of a_k squares of the kth mosaic, and a figure H_k, made up of b_k squares of the mosaic, satisfying the conditions $G_k \subset F \subset H_k$, $b_k/10^{2k} - a_k/10^{2k} < \varepsilon$. Since $s(H_k) - s(G_k) = b_k/10^{2k} - a_k/10^{2k} < \varepsilon$, the polygons G_k and H_k are the ones we are looking for.

Conversely, suppose that for every $\varepsilon > 0$ there exist polygons G and H such that $s(H) - s(G) < \varepsilon$ and $G \subset F \subset H$. For the same $\varepsilon > 0$ we can make k so large that there exist figures G_ε and H_ε, made up of squares of the kth mosaic and satisfying the conditions $G_\varepsilon \subset G$, $H_\varepsilon \supset H$, $s(G) - s(G_\varepsilon) < \varepsilon$, $s(H_\varepsilon) - s(H) < \varepsilon$. Thus $G_\varepsilon \subset F \subset H_\varepsilon$, with $s(H_\varepsilon) - s(G_\varepsilon) < 3\varepsilon$, i.e., $b_k/10^{2k} - a_k/10^{2k} < 3\varepsilon$, where a_k and b_k are the numbers of squares of the kth mosaic making up the figures G_ε and H_ε, respectively. Taking the limit (as $\varepsilon \to 0$), we get $\bar{s}(F) - \underline{s}(F) \leqslant 0$, i.e., $\bar{s}(F) = \underline{s}(F)$, so that the figure F is measurable.

Theorem 4. *If the figures F' and F" are measurable, then so are the figures* $F' \cup F''$, $F' \cap F''$ *and* $F' \setminus F''$.

PROOF. Choose an arbitrary $\varepsilon > 0$, and let G', G'', H', H'' be polygons such that $G' \subset F' \subset H'$, $G'' \subset F'' \subset H''$ and $s(H') - s(G') < \varepsilon/4$, $s(H'') - s(G'') < \varepsilon/4$ (see Fig. 9; these polygons exist because of Theorem 3). We can also assume (enlarging the polygon H'' somewhat, if necessary) that $F'' \subset \operatorname{int} H''$ where int denotes the interior of the polygon in question. Then

$$G' \setminus \operatorname{int} H'' \subset F' \setminus F'' \subset$$
$$\subset H' \setminus \operatorname{int} G'',$$

---

[3] There is a paucity of results in the theory of equidecomposability for *arbitrary* sets; see, for example, [28], Chap. III.

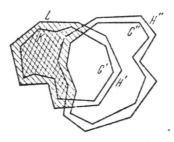

FIG. 9

i.e., $K = G' \setminus$ int H'', $L = H' \setminus$ int G'' are polygons satisfying the condition $K \subset F' \setminus F'' \subset L$. We now show that $s(L) - s(K) < \varepsilon$, thereby proving the measurability of the set $F' \setminus F''$. It follows from axiom (β) that $s(\overline{H' \setminus G'}) = s(H') - s(G') < \varepsilon/4$, and similarly $s(\overline{H'' \setminus G''}) < \varepsilon/4$. Hence, for sufficiently large k there exist figures Q_1 and Q_2, made up of squares from the kth mosaic, such that

$$\overline{H' \setminus G'} \subset Q_1, \quad \overline{H'' \setminus G''} \subset Q_2, \quad s(Q_1) < \varepsilon/2, \quad s(Q_2) < \varepsilon/2.$$

It follows from the easily verified inclusion relation

$$\overline{L \setminus K} \subset \overline{H' \setminus G'} \cup \overline{H'' \setminus G''}$$

(see Fig. 9) that $\overline{L \setminus K}$ is contained in the figure $Q_1 \cup Q_2$, made up of squares from the kth mosaic and satisfying the condition $s(Q_1 \cup Q_2) \leqslant s(Q_1) + s(Q_2) < \varepsilon$. Therefore $s(\overline{L \setminus K}) < \varepsilon$, i.e., $s(L) - s(K) < \varepsilon$, and this implies the measurability of the set $F' \setminus F''$.

Now let M be a polygon containing both figures F' and F''. By what was just proved, it follows from the formulas

$$F' \cup F'' = M \setminus (M \setminus (F' \cup F'')) = M \setminus ((M \setminus F') \setminus F''),$$

$$F' \cap F'' = M \setminus (M \setminus (F' \cap F'')) = M \setminus ((M \setminus F') \cup (M \setminus F''))$$

that each of the figures $F' \cup F''$ and $F' \cap F''$ is also measurable.

Theorem 5. *There exists one and only one function s, defined on the set of all measurable figures, which satisfies the conditions* (α), (β), (γ), (δ).

The proof is analogous to that of Theorem 2. The area $s(F)$ of every measurable figure F is defined, and the function s satisfies the conditions (α), (β), (γ), (δ). This proves the existence of s.

To prove the uniqueness of s, let s^* be any other function which is defined on all measurable figures and satisfies the conditions (α), (β), (γ), (δ). Given any measurable figure F and any $\varepsilon > 0$, we choose polygons G and H such that $G \subset F \subset H$ and $s(H) - s(G) < \varepsilon$. By (α) and (β) we have

$$s^* (F) = s^* (G + (F \setminus G)) = s^* (G) + s^* (F \setminus G) \geqslant s^* (G).$$

(bearing in mind that the figure $F\backslash G$ is measurable, by Theorem 4), and in just the same way $s^* (F) \leqslant s^* (H)$. Moreover, $s^* (G) = s (G)$, $s^* (H) = s (H)$, by Theorem 2, and hence

$$s (G) \leqslant s^* (F) \leqslant s (H).$$

The analogous double inequality also holds for $s(F)$:

$$s (G) \leqslant s (F) \leqslant s (H).$$

It follows from these inequalities that $|s^* (F) - s (F)| \leqslant s (H) - s (G) < \varepsilon$. Since $\varepsilon > 0$ is arbitrary, we have $|s^* (F) - s (F)| = 0$, i.e., $s^* (F) = s (F)$, and the uniqueness is proved.

§3. Further properties of area

Since axioms (α), (β), (γ), (δ) uniquely define the concept of area, it is natural to expect that all further properties of area can be derived by using these axioms *exclusively*, and in fact we now use the axiomatic approach to derive some more properties of area. Two of these properties will be denoted by (α^*) and (γ^*), since their meaning is close to that of the previously formulated axioms (α) and (γ).

(α^*) *If two measurable figures F and G obey the inclusion relation $F \subset H$, then $s (F) \leqslant s (H)$.*

Let F and H be polygons, and let G be the polygon such that $H = F + G$, i.e., $G = \overline{H \setminus F}$. By axiom (β) we have $s (H) = s (F) + s (G)$. and moreover $s (G) \geqslant 0$ (by axiom (α)). Therefore, $s (H) \geqslant s (F)$.

In defining the figure G, we used the operation of closure, because of our convention that all polygons are regarded as *closed*. The proof is simpler for arbitrary measurable figures F and H:

$$s\,(H) = s\,(F) + s\,(H \setminus F) \geqslant s\,(F).$$

(see Theorem 4).

The property (α^*) just proved (the *monotonicity* of area) is often taken as an axiom replacing (α), i.e., a theory of area is constructed based on the axioms (α^*), (β), (γ), (δ). (In fact, this set of axioms is equivalent to the previous set, since the validity of (α) is implied by axioms (β), (γ), (δ) and the monotonicity property.)

Theorem 6. *The formula*

$$s\,(F' \cup F'') = s\,(F') + s\,(F'') - s\,(F' \cap F'')$$

holds for arbitrary measurable figures F' and F''.

PROOF. By axiom (β), we have

$$s\,(F' \cup F'') = s\,(F') + s\,(F'' \setminus F')$$

since $F' \cup F'' = F' + (F'' \setminus F')$ (see also Theorem 4), and similarly

$$s\,(F'') = s\,(F'' \setminus F') + s\,(F' \cap F'')$$

since $F'' = (F'' \setminus F') + (F' \cap F'')$, The theorem now follows from these two formulas taken together.

Corollary. *Given any two measurable figures F' and F'',*

$$s\,(F' \cup F'') \leqslant s\,(F') + s\,(F'')$$

To establish further properties of area, we will need the well known formula for the area of a rectangle.

Theorem 7. *The area of a rectangle F can be calculated from the formula $s\,(F) = ab$, where a and b are the lengths of the sides of F.*

The *proof* will be given first for the case where the sides of the rectangle F are parallel to the sides of the unit square Q. Then from the validity of the theorem in this special case we will deduce another property of area, after which the validity of Theorem 7 in the general case will be obvious.

Thus, let the sides of the rectangle F be parallel to the sides of the unit square Q (i.e., to the lines of the mosaic). By property (γ) we can also assume (making a translation, if necessary) that the lower left-hand vertex of the rectangle F coincides with one of the vertices of the zeroth mosaic. Let p_0 denote the vertex of the kth mosaic which is contained in the rectangle F and is closest to the upper right-hand vertex of F (see Fig. 10). Then the figure made up of all the squares of the kth mosaic contained in F is a rectangle G_k, with the points p_0 and q_0 as opposite vertices. Let α_k and β_k be the numbers of squares of the kth mosaic adjacent to the sides of the rectangle G_k Then

$$a \cdot 10^k - 1 < \alpha_k \leqslant a \cdot 10^k,$$
$$b \cdot 10^k - 1 < \beta_k \leqslant b \cdot 10^k.$$

Hence the number of squares of the kth mosaic making up G_k, i.e., the number $a_k = \alpha_k \beta_k$, satisfies the inequalities

$$(a \cdot 10^k - 1)(b \cdot 10^k - 1) < a_k \leqslant ab \cdot 10^{2k}.$$

Bordering G_k from the right and from above with a strip of width equal to one square of the kth mosaic, we get a rectangle $H_k \supset F$

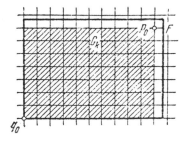

FIG. 10

made up of $b_k = (\alpha_k + 1)(\beta_k + 1)$ squares of the kth mosaic, and hence

$$ab \cdot 10^{2k} < b_k \leqslant (a \cdot 10^k + 1)(b \cdot 10^k + 1).$$

It follows from these inequalities that

$$\frac{a_k}{10^{2k}} > \left(a - \frac{1}{10^k}\right)\left(b - \frac{1}{10^k}\right),$$

$$\frac{b_k}{10^{2k}} \leqslant \left(a + \frac{1}{10^k}\right)\left(b + \frac{1}{10^k}\right),$$

from which, taking the limit as $k \to \infty$ (see (1), (2)), we find that $\underline{s}(F) \geqslant ab$, $\bar{s}(F) \leqslant ab$. Therefore $\underline{s}(F) = \bar{s}(F) = ab$, i.e., $s(F) = ab$.

Thus Theorem 7 holds if the sides of the rectangle are parallel to the sides of the unit square Q. From this we can deduce the following property:

(γ^*) *Area is invariant under arbitrary motions, i.e., if F is a measurable figure and $F' = g(F)$ is the figure obtained by subjecting F to any motion g, then the figure F' is measurable and $s(F') = s(F)$.*

PROOF. Let $Q' = g(Q)$, where Q is the unit square, and let P be the square inscribed about Q', whose sides are parallel to the sides of the square Q (Fig. 11a). The square P decomposes into the square Q' and four congruent right triangles T_1, T_2, T_3, T_4, the lengths of whose legs we denote by a and b. By the Pythagorean theorem, $a^2 + b^2 = 1$ (since the hypotenuse of the triangle T_1, i.e., the side of the square Q', is of length 1). In Fig. 11b we show the decomposition of the square P into two squares Q_1, Q_2 with sides a, b, and four triangles T'_1, T'_2, T'_3, T'_4. By axiom (β),

$$s(P) = s(Q') + s(T_1) + s(T_2) + s(T_3) + s(T_4),$$

$$s(P) = s(Q_1) + s(Q_2) + s(T'_1) + s(T'_2) + s(T'_3) + s(T'_4).$$

and moreover $s(T_i) = s(T'_i)$, $i = 1, 2, 3, 4$, since T'_i is obtained by subjecting T_i to a translation (axiom (γ)). Therefore

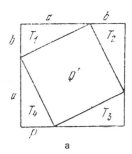
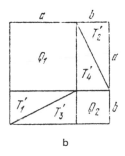

a   b

FIG. 11

$$s (Q') = s (Q_1) + s (Q_2).$$

But the sides of the squares Q_1 and Q_2 are *parallel* to the sides of the unit square Q, and hence, by Theorem 7, $s (Q_1) = a^2$, $s (Q_2) = b^2$, i.e., $s (Q') = a^2 + b^2 = 1$. Thus *every* square Q' congruent to the unit square Q is of area 1.

Now let $s^* (F) = s (g (F))$ for an arbitrary polygon F. It is clear that the function s^* satisfies axiom (α). Moreover, if the polygons F and F' can be obtained from one another with the help of a translation, then so can $g (F)$ and $g (F')$, and hence $s^* (F) = s^* (F')$, i.e., the function s^* also satisfies axiom (γ). Axiom (β) can be verified just as easily. Finally, $s^* (Q) = s (g (Q)) = s (Q') = 1$ (as proved above), i.e., s^* satisfies axiom (δ). Thus s^* satisfies all the axioms (α), (β), (γ), (δ). Therefore, by Theorem 2, the function s^* *coincides with* s, i.e., $s^* (F) = s (F)$ for every polygon F. In other words $s (g (F)) = s (F)$.

Thus we have proved property (γ^*) for polygons. This immediately implies the validity of Theorem 7 for *arbitrary* rectangles.

Now let F be an arbitrary measurable figure, and let $\varepsilon > 0$. We choose polygons G and H such that $G \subset F \subset H$ and $s (H) - s (G) < \varepsilon$ (Theorem 3). Then the polygons $G' = g (G)$, and $H' = g (H)$ satisfy the condition

$G' \subset g(F) \subset H'$, where, as just shown, $s(G') = s(G), s(H') = s(H)$, and therefore $s(H') - s(G') = s(H) - s(G) < \varepsilon$. It follows that the figure $F' = g(F)$ is measurable. The validity of property (γ^*) for arbitrary measurable figures is now proved in just the same way as for rectangles (with a reference to Theorem 5 rather than to Theorem 2).

The statement (γ^*) is a more general property of area than axiom (γ), and it is sometimes chosen as an axiom instead of (γ), leading to the construction of a theory of area based on the properties (α), (β), (γ^*), (δ), regarded as axioms. The above considerations show that this set of axioms is equivalent to the original set. We will subsequently often have just this set of axioms $(\alpha), (\beta), (\gamma^*), (\delta)$ in mind.

Theorem 8. *The area of figures is invariant under replacement of the unit square Q with a square congruent to it.*

PROOF. Let Q^* be a square congruent to Q, and let s be the area constructed by using Q as the unit square, while s^* is the area constructed by using Q^*. By Theorem 7, $s^*(Q) = s^*(Q^*) = 1$. Hence the functions s and s^*, considered on the set of all polygons, satisfy all the axioms (α), (β), (γ), (δ), (with Q as the unit square), and therefore, by Theorem 2, $s(F) = s^*(F)$ for an arbitrary polygon F.

Theorem 3 now shows that the class of measurable figures is one and the same for both s and s^*. But then, by Theorem 5, $s(F) = s^*(F)$ for an arbitrary measurable figure F.

It follows from Theorem 8 that if the zeroth mosaic is replaced by any congruent mosaic in the constructive definition of area (§1), then the area of figures (as well as the class of measurable figures) *does not change.*

Theorem 9. *Let f be a similarity transformation with ratio k, i.e., a mapping of the plane onto the same plane which magnifies distances k times. and let F be a measurable figure. Then the figure $F' = f(F)$ is also measurable and $s(F') = k^2 s(F)$.*

PROOF. Let $s^*(F) = \dfrac{1}{k^2} s(f(F))$. Since f carries every polygon into another polygon, the function s^* is defined on the set of all polygons. It is easily verified that s^* satisfies axioms (α), (β), (γ). Moreover, if Q is the unit square, then $f(Q)$ is a square with sides of length k. By Theorem 7, $s(f(Q)) = k^2$, i.e., $s^*(Q) = \frac{1}{k^2} s(f(Q)) = 1$, so that s^* also satisfies axiom (δ). It follows from Theorem 2 that $s^*(F) = s(F)$ for every polygon F, i.e., $s(f(F)) = k^2 s(F)$.

This implies that if F is a measurable figure, then so is $f(F)$ (Theorem 3). Hence the function s^* is defined on the set of all measurable figures. Since s^* satisfies axioms (α), (β), (γ), (δ), we have $s^*(F) = s(F)$ for every measurable figure, i.e., $s(f(F)) = k^2 s(F)$.

§4. The independence of the area axioms

As we now show, the axioms (α), (β), (γ), (δ) are *independent*. In other words, each of the axioms is essential; i.e., if one of the axioms is dropped, we can construct a function on the set of measurable figures which satisfies the remaining axioms and is different from the area.

It is easiest of all to establish the independence of axiom (δ). In fact, setting $s_\delta(F) = 0$ for every measurable figure F, we get a function s_δ different from the area s, which satisfies axioms (α), (β), (γ). As another example, let $s_\delta(F) = \lambda s(F)$, where $\lambda > 0$ is a fixed number. This function also satisfies all the axioms (α), (β), (γ), but it does not satisfy axiom (δ) (if $\lambda \neq 1$). It can be shown (by the same method as used to prove Theorem 9) that *every* function satisfying axioms (α) (β), (γ) is of the form $\lambda s(F)$, where $\lambda \geq 0$.

It is also an easy matter to construct a function showing the independence of axiom (β). In fact, we need only set $s_\beta(F) = 1$ for every measurable figure F (the fact that axioms (α), (γ), (δ) hold in this case is obvious). As another example, consider the function $s_\beta(F) = (s(F))^2$ or $s_\beta(F) = e^{s(F)-1}$.

To construct an example showing the independence of axiom (γ), let P_1 and P_2 denote the two half-planes into which the plane is divided by a fixed line l, where it is assumed that the unit square Q lies in the half-plane P_1. Given any measurable figure F, we set

$$s_\gamma (F) = s (F \cap P_1) + 2s (F \cap P_2).$$

It is easy to see that the function s_γ satisfies axioms (α), (β), (δ). At the same time, axiom (γ) is not satisfied. For example, if Q' is a square obtained by subjecting Q to a translation and if Q' lies in the half-plane P_2, then $s_\gamma(Q') = 2$, $s_\gamma (Q) = 1$, i.e., $s_\gamma (Q') \neq s_\gamma (Q)$. We can also define the function s_γ differently, as

$$s_\gamma (F) = s (F \setminus H) + \lambda s (F \cap H),$$

where H is a fixed measurable figure of positive area which does not intersect Q, and $\lambda \neq 1$ is a fixed positive number.

Finally, we turn to the problem of the independence of axiom (α). The problem has two aspects, depending on whether we consider the set of axioms (α), (β), (γ), (δ) or the set (α), (β), (γ^*), (δ). The problem of showing that axiom (α) is independent of (β), (γ^*), (δ) is more complicated but more fundamental. We are talking about constructing a function $s_\alpha (F)$ which fails to satisfy axiom (α), at the same time as it satisfies axioms (β), (δ) and is invariant under *arbitrary* motions of the plane. In particular, the function $s_\alpha (F)$ will be invariant under translations and thus the proof that axiom (α) is independent of (β), (γ^*), (δ) *contains* the proof that axiom (α) is independent of axioms (β), (γ), (δ). It is just this aspect (axiom (α)'s independence of axioms (β), (γ^*), (δ)) which will be considered here. The simpler problem of showing that axiom (α) is independent of (β), (γ), (δ) is considered in §10.

Thus we must construct a function $s_\alpha (F)$ which satisfies axioms (β), (γ^*), (δ), but does not have the property (α). The construction of $s_\alpha (F)$ involves delicate questions of set theory and rests on the *axiom of choice* (or on some equivalent propositions, for example, *Zorn's lemma*). Let $f(x)$ be an *additive function* (defined for real x and taking real values), i.e., a function such that

$$f (x' + x'') = f (x') + f (x'') \tag{3}$$

for arbitrary x', x''. Setting $s_\alpha (F) = f (s (F))$, where s is the ordinary area, we get a function s_α which is defined on the set of all measurable figures and satisfies axioms (β), (γ^*). For example, if $F = F' + F''$, then, by (3) and axiom (β) for the area s, we have

$$s_\alpha (F' + F'') = f (s (F' + F'')) =$$
$$= f (s (F') + s (F'')) = f (s (F')) + f (s (F'')) =$$
$$= s_\alpha (F') + s_\alpha (F''),$$

i.e., s_α satisfies axiom (β). It is easy to see that s_α also satisfies axiom (γ^*). If, in addition, $f(x)$ satisfies the condition

$$f (1) = 1, \tag{4}$$

then the function s_α also satisfies axiom (δ), since

$$s_\alpha (Q) = f (s (Q)) = f (1) = 1.$$

Thus the required function $s_\alpha (F) = f (s (F))$ will be constructed, once we succeed in constructing an additive function $f(x)$ satisfying condition (4) and taking a negative value for some $x > 0$. For example, suppose the function $f(x)$ takes the value $f (\sqrt{5}) = -1$ for $x = \sqrt{5}$. Then, by Theorem 7, for the rectangle F with sides of length 1 and $\sqrt{5}$ we have

$$s_\alpha (F) = f (s (F)) = f (1 \cdot \sqrt{5}) = f (\sqrt{5}) = -1,$$

so that axiom (α) is not satisfied.

Every linear function $f (x) = \lambda x$ is additive (satisfies condition (3)). But if $f(x)$ satisfies condition (4), then $\lambda \cdot 1 = 1$, i.e., $\lambda = 1$, and the resulting function $f (x) = x$ takes a *positive* value for every $x > 0$. Thus the required additive function cannot be linear.

We now study the properties of additive functions. From (3) we find that

$$f (2x) = f (x + x) = f (x) + f (x) = 2f (x),$$

$$f(3x) = f(2x + x) = f(2x) + f(x)$$
$$= 2f(x) + f(x) = 3f(x)$$

and so on. Thus we have

$$f(rx) = rf(x). \tag{5}$$

for every natural number r. Moreover, given natural numbers m and n, we deduce from (3) and (5) that

$$mf(x) = f(mx) = f((m - n)x + nx) =$$
$$= f((m - n)x) + f(nx) = f((m - n)x) + nf(x),$$

i.e., $f((m - n)x) = (m - n)f(x)$. This means that formula (5) is also valid for $r = m - n$, i.e., for arbitrary *integers* r. Finally, if p and $q \neq 0$ are integers, then

$$pf(x) = f(px) = f\left(q \cdot \left(\frac{p}{q}x\right)\right) = qf\left(\frac{p}{q}x\right),$$

and hence $f\left(\frac{p}{q}x\right) = \frac{p}{q}f(x)$. Thus formula (5) is valid for every *rational* number r. We also observe that (3) and (5) imply the formula

$$f(r_1 x_1 + \ldots + r_k x_k) = r_1 f(x_1) + \ldots + r_k f(x_k), \tag{6}$$

valid for arbitrary rational $r_1, \ldots, r_k$ and real $x_1, \ldots, x_k$.

If the additive function $f(x)$ is *continuous*, then the validity of formula (5) for *rational* r implies its validity "by continuity" for arbitrary *real* r. In particular, for $x = 1$ we get $f(r) = rf(1)$, i.e., $f(r) = \lambda r$ where $\lambda = f(1)$. Since this holds for arbitrary real r, the function $f(x)$ is linear. (It can be shown that a measurable additive function is also necessarily linear.) Thus the additive function that we need cannot be linear (and, moreover, it must be non-measurable). We now show that such a function can be constructed.

Let $\xi_1 = 1$. According to (4), $f(\xi_1) = f(1) = 1$. Now choose any other number ξ_2. If ξ_2 is rational, the value of $f(\xi_2$ is uniquely determined, namely $f(\xi_2) = f(\xi_2 \cdot 1) = \xi_2 f(1) = \xi_2$. However, if ξ_2 is irrational, say $\xi_2 = \sqrt{5}$, then formula (5) or (6)) does not give us the value of $f(\xi_2)$, and we can choose it arbitrarily, setting $f(\sqrt{5}) = -1$, say. Next we choose another number ξ_3. If ξ_3 is a linear combination with rational coefficients of the numbers ξ_1 and ξ_2, for example, if $\xi_3 = {}^3/_5 \cdot 1 - 7\sqrt{5}$, then, by (6), $f(\xi_3) = {}^3/_5 f(1) - 7f(\sqrt{5}) = {}^3/_5 \cdot 1 - 7 \cdot (-1) = {}^{38}/_5$. However, if ξ_3 is not a rational linear combination of the numbers ξ_1 and ξ_2, then formula (6) does not give us the value of $f(\xi_3)$, and we can choose it arbitrarily.

Suppose the function $f(x)$ is already defined on some (finite or infinite) set $A = \{\xi_1, \xi_2, \xi_3, \ldots\}$, and choose yet another number ξ_α. If ξ_α can be represented as a rational linear combination of a finite set of numbers from A, then formula (6) allows us to determine the value of $f(\xi_\alpha)$ uniquely. However, if ξ_α cannot be represented as such a combination, we can choose $f(\xi_\alpha)$ arbitrarily, thereby defining the function $f(x)$ at the point $x = \xi_\alpha$, i.e., we can adjoin the point ξ_α to the set A, on which the function $f(x)$ is already defined. We can then adjoin more and more points to A.

Is it possible, by choosing more and more new numbers, to "sort through" the whole set R of real numbers? If so, then we obtain the required function as a result. Contemporary mathematics (more exactly, a theory of sets in which the *axiom of choice* is satisfied) allows us to consider constructions that "sort through" all the elements of an arbitrary set (in particular, the set R). This means that the required additive function $f(x)$ exists (in the context of a set theory allowing the axiom of choice). The existence of a function $s_\alpha(F) = f(s(F))$ confirming the independence of axiom (α) is now proved.

Discontinuous additive functions were first considered in the work of Hamel [31], and are called *Cauchy–Hamel functions*. Our way of constructing such functions can be presented somewhat differently, by using the concept of a *rational basis* for the number line R. A set $B \subset R$ is called a rational basis if it has the following two properties:

a) Every number $x \in R$ can be represented as a rational linear combination of numbers of R, i.e., there exists a positive integer m, numbers $b_1, \ldots, b_m \in B$ and rational numbers $r_1, \ldots, r_m$ such that $x = r_1 b_1 + \ldots + r_m b_m$;

b) No number $b \in B$ can be represented as a rational linear combination of numbers of the set B which are not equal to b.

The argument given above, which "sorts through" the set R and leads to the construction of the Cauchy–Hamel function $f(x)$, is actually tantamount to the construction of a rational basis for the real line. In fact, suppose we define the set B by assigning to it the numbers $\xi_1 = 1$, $\xi_2 = \sqrt{5}$, and also every number ξ_α which cannot be represented as a rational linear combination of numbers previously encountered in the sorting process. Then it is easy to see that the set B is a rational basis for the number line R.

The construction of the Cauchy–Hamel function just described can be thought of as follows: The values of the function $f(x)$ are chosen *arbitrarily* at the points of the basis B, after which the values of $f(x)$ at the remaining points (which are rational combinations of the basis elements) are uniquely determined by formula (6). From this it is clear that there exist 2^x Cauchy-Hamel functions. To prove the independence of axiom (α), we need *one* such function. For example, we can set $f(1) = 1$, and $f(x) = -1$ for all the other elements of the basis (in particular, $f(\sqrt{5}) = -1$), afterwards using (6) to extend $f(x)$ onto the whole set R.

Finally, we give a complete proof of the existence of a rational basis B containing the numbers $\xi_1 = 1, \xi_2 = \sqrt{5}$ (thereby proving the existence of the required Cauchy–Hamel function). To do so, we will use not the axiom of choice itself, but rather an equivalent (and now more commonly used) proposition, known as *Zorn's lemma* [58].

A set M is said to be *ordered* if a rule is given specifying which pairs (a, b) of distinct elements of M satisfy the relation $a < b$. Here it is required that the relations $a < b$, $b < c$ imply the relation $a < c$, and that there be no elements $a, b \in M$ such that the relations $a < b$, $b < a$ hold simultaneously. If neither of the relations $a < b, b < a$ holds for given elements $a, b \in M$ $(a \neq b)$, then a

and b are said to be *noncomparable*. A subset $C \subset M$ is called a *chain* if it does not contain a pair of noncomparable elements, i.e., if given any a, $b \in C$ $(a \neq b)$, either $a < b$ or $b < a$. A chain $C \subset M$ is said to be *bounded* if there exists an element $q \in M$ (either belonging to C or not) such that $a \leqslant q$ for every element $a \in C$. An element $m \in M$ is said to be *maximal* if there does not exist an element $a \in M$ satisfying the condition $m < a$. It is clear that if the ordered set M is itself a chain, then it can have no more than *one* maximal element. However, in an arbitrary ordered set there can exist more than one maximal element.

Zorn's lemma. *If all the chains contained in an ordered set M are bounded, then M contains at least one maximal element.*

Zorn's lemma represents one form of the axiom of choice. It is now known (see, for example, [57]) that a theory of sets including the axiom of choice is *noncontradictory* (the same is true of a theory of sets rejecting the axiom, i.e., including its negation).

To prove the existence of a rational basis for the number line R, we agree to call a subset $B \subset R$ *rationally independent* if it satisfies the above-mentioned condition b) (figuring in the definition of a rational basis). Let M denote the set of all rationally independent subsets $B \subset R$ which contain the numbers 1 and $\sqrt{5}$. Given two elements B_1, $B_2 \in M$ (i.e., two rationally independent subsets $B_1 \subset R$, $B_2 \subset R$), we will take $B_1 < B_2$ to mean that $B_1 \subset B_2$.

Now given any chain $C \subset M$, let B_C denote the union of all the elements $B \in C$ (considered as a subset of the number line R). It is easy to see that the set B_C is rationally independent, i.e., $B_C \in M$. In fact, let $x_1, \ldots, x_m$ be arbitrary (distinct) numbers belonging to the set B_C. For every $i = 1, \ldots, m$ we can find a $B_i \in C$, such that $x_i \in B_i$. Since C is a chain, then among the finite number of elements $B_1, \ldots, B_m$ there is a largest element; to be explicit, let this largest element be B_m, so that $B_i \subset B_m$, $i = 1, \ldots, m$. Then $x_i \in B_i \subset B_m$, i.e., all the numbers $x_1, \ldots, x_m$ are contained in B_m, and hence are rationally independent, since $B_m \in M$. Moreover, since $B < B_C$ for every $B \in C$, by definition, the chain C is bounded.

Thus every chain $C \subset M$ is bounded. By Zorn's lemma, there exists a *maximal* element B^* in M. In other words, M^* is a rationally independent subset of the number line R such that no *larger* subset is rationally independent. This means that every real number x can be represented as a linear combination (with rational coefficients) of numbers in B^*. Thus B^* satisfies condition a), i.e., B^* is a rational basis (containing the numbers 1 and $\sqrt{5}$). Our proof is now complete.

We now make some concluding remarks on the subject of the independence of the axioms (α), (β), (γ^*), (δ) defining the concept of area. The independence of axiom (α) means, in particular, that *the formula for the area of a rectangle cannot be deduced from axioms (β), (γ^*), (δ) alone without using axiom (α).* In fact, the function $s_\alpha (F) = f(s(F))$ constructed above satisfies axioms (β), (γ^*), (δ) and hence any argument using only the axioms (β), (γ^*), (δ) applies just as well to the function $s_\alpha (F)$ as to the area $s(F)$. Hence, if the formula for the area of a rectangle could be derived from axioms (β), (γ^*), (δ) alone, then the formula would also be valid for the function $s_\alpha (F)$. But the function s_α takes the value $s_\alpha (P) = -1$ for a rectangle P whose sides are of length 1 and $\sqrt{5}$, which is *different* from the value $s(P) = 1 \cdot \sqrt{5}$ obtained by using the formula for the area of a rectangle.

On the other hand, in a theory of sets rejecting the axiom of choice, there *does not exist* a function $s_\alpha (F)$ which satisfies the axioms (β), (γ^*), (δ) and is different from the area. For otherwise, letting P_x denote the rectangle with sides of length 1 and x, and then letting $f(x) = s_\alpha (P_x)$, we would get a Cauchy-Hamel function, i.e., a nonmeasurable function. But then the axiom of choice would hold, by the results of reference [47]. In other words, in a theory of sets rejecting the axiom of choice, the only function satisfying axioms (β), (γ^*), (δ), alone is the area $s(F)$. Moreover, if the formula for the area of a rectangle can be derived in a given set theory by using axioms (β), (γ^*), (δ) alone, then the derivation must necessarily rest on the negation of the axiom of choice.

Suppose that arguments using facts depending on whether or not the axiom of choice holds are not regarded as part of "elementary geometry." Then we are compelled to admit that the formula for the area of a rectangle cannot be derived from axioms (β), (γ^*), (δ) alone without using axiom (α) (since it would have to be possible to carry out such a derivation both in a theory accepting the axiom of choice and in a theory rejecting that axiom). In other words, from the standpoint of "elementary geometry," axiom (α) must be regarded as independent of axioms (β), (γ^*), (δ).

We emphasize once more that the problem proving that axiom (α) is independent of axioms (β), (γ), (δ) is simpler, and can be solved without using the axiom of choice (see the footnote on p. 81).

§5. Methods for calculating the area of figures

Axiom (α) (or the equivalent property of monotonicity of area) can be used directly only to obtain *estimates* (inequalities). But from these inequalities we can deduce *equalities* (i.e., exact values of the areas of figures) by passing to the limit. The general pattern of this kind of argument, involving a passage to the limit, can be formulated as follows:

Given a measurable figure F, consider a sequence $G_1, G_2, \ldots$ of measurable figures imbedded in F. If the area of the part of the figure F not occupied by the figure G_k decreases without limit as $k \to \infty$, then $s(F) = \lim\limits_{k \to \infty} s(G_k)$.

The proof is an immediate consequence of the formulas $s(F) = s(G_k) + s(F \setminus G_k)$ and $\lim\limits_{k \to \infty} s(F \setminus G_k) = 0$. This method of calculating the area of the figure F is called the *method of exhaustion*, since the figures $G_1, G_2, \ldots$ progressively "exhaust" the entire area of the figure F. There is also a modification of the method of exhaustion, in which one considers a sequence $H_1, H_2, \ldots$ of enveloping figures (containing F), in addition to the sequence $G_1, G_2, \ldots$ of imbedded figures. If, in this case, $\lim\limits_{k \to \infty} s(H_k \setminus G_k) = 0$, then the formula $\lim\limits_{k \to \infty} s(G_k) = s(F)$ holds, as before. In fact, since $F \setminus G_k \subset H_k \setminus G_k$, it follows from axiom (α) that $0 \leqslant s(F \setminus G_k) \leqslant s(H_k \setminus G_k)$, and hence $\lim\limits_{k \to \infty} s(F \setminus G_k) = 0$, which allows us to apply the method of exhaustion in its original formulation. The enveloping figures H_k are used as a convenient way of *estimating* the area of the figure $F \setminus G_k$ (since it may turn out that it is easier to calculate the area of the figure $H_k \setminus G_k$ than that of the figure $F \setminus G_k$). There is still another modification of the method of exhaustion; namely, to prove the equality $\lim_{k \to \infty} s(H_k \setminus G_k) = 0$, we need only verify that $\lim_{k \to \infty} s(G_k) = \lim_{k \to \infty} s(H_k)$ (or even that $\lim_{k \to \infty} s(G_k) \geqslant \lim_{k \to \infty} s(H_k)$, since the opposite inequality is obvious).

We have already encountered an example of the method of exhaustion in §3, in calculating the area of a rectangle. In this case, G_k and H_k were taken to be rectangles made up of squares from the kth mosaic, with $G_k \subset F \subset H_k$, where F is the given rectangle (see Fig. 10). Here the figures G_k and H_k are much *simpler* than F, since they are made up of squares from the kth mosaic and hence have areas that can be calculated directly. Moreover, since

$$s(G_k) > \left(a - \frac{1}{10^k} \right)\left(b - \frac{1}{10^k} \right),$$

$$s(H_k) \leqslant \left(a + \frac{1}{10^k} \right)\left(b + \frac{1}{10^k} \right),$$

we have $\lim_{k \to \infty} s(G_k) \geqslant \lim_{k \to \infty} s(H_k)$, from which it follows that

$$s(F) = \lim_{k \to \infty} s(G_k) = \lim_{k \to \infty} s(H_k) = ab.$$

In deriving this formula, essential use is made of both axiom (α) and passage to the limit (i.e., the "method of exhaustion").

Another example, familiar from high school, is the use of the method of exhaustion to calculate the area of a circular disk,[4] In

---

[4] If G_k is the regular 2^k-gon inscribed in the disk F, and H_k the regular 2^k-gon circumscribed about F, then, as simple calculations show, the "band" between H_k and G_k decreases indefinitely in area as $k \to \infty$, i.e., $\lim_{k \to \infty} s(H_k/G_k) = 0$, and therefore $s(F) = \lim_{k \to \infty} s(G_k)$. However, this last formula is usually not proved in high school, but rather is taken as the "definition" of the area of the disk F.

deriving the formula

$$s\,(F) = \int_a^b f\,(x)\,dx$$

for the area of the "curvilinear trapezoid" bounded by the x-axis, the vertical lines $x = a$, $x = b$ and the graph of a positive function $y = f\,(x)$, it is also easy to observe the operation of the method of exhaustion. In fact, here G_h and H_h are "step-shaped figures" (see Fig. 12).

Thus the method of exhaustion involves the use of a limiting process, and, in essence, consists of integration in either explicit or veiled form. The notion of a limit and the associated "ε-technique" are among the more complicated and nonelementary concepts of undergraduate mathematics. For this reason, we will classify the method of exhaustion (or every argument using axiom (α) and hence equivalent, in this sense, to the method of exhaustion) as a *nonelementary* method of calculating area. On the other hand, we will regard arguments and methods based only on the use of axioms (β), (γ), (δ) (or (β), (γ^*), (δ))) and, of course, not involving the axiom of choice, as *elementary* methods of calculating area. From this point of view, the theorems on the area of the rectangle and the circular disk are *nonelementary*. And in fact, among the matters treated in the elementary theory of area, it is just these theorems (when

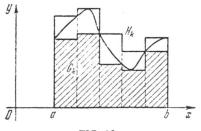

FIG. 12

equipped with full and accurate proofs) which are the most complicated and difficult to understand.

Before turning to the characteristics of elementary methods of calculating area, we discuss the connection between the theory of area presented above and the typical course in high school geometry. The high school concept of area is very close to the axiomatic concept of area (see §2). In fact, with the axiomatic approach there is no need for a *definition* of area, i.e., roughly speaking, area is understood to be "that which satisfies axioms (α), (β), (γ^*), (δ) (which, admittedly, are not called axioms, but "properties" of area, and are regarded as obvious). In any event, the assertions (α), (β), (γ^*), (δ) are familiar to every high school student, although perhaps in a somewhat different form. For example, axiom (γ^*) is known as the statement that "congruent figures have the same area."

In the axiomatic approach we have an *existence and uniqueness theorem*, which greatly enriches our idea of area. The existence guarantees that the set of axioms for area are noncontradictory. The uniqueness is no less significant, and is a convenient tool for proving theorems about properties of area (as we saw in §3). However, one might say that the high school student is already convinced of the existence and uniqueness of area (i.e., of "that which has properties (α), (β), (γ^*), (δ)"). In fact, area is an abstraction reflecting objective properties of real objects, properties about which students already have a good idea, both as a result of everyday experience and early encounters with the concept of area in their elementary mathematics and physics lessons. The objectivity of this concept *replaces* the existence theorem, and leads the students to believe that they are studying something which actually exists. As for the uniqueness, the fact that, in the last analysis, properties (α), (β), (γ^*), (δ) can be used to *uniquely* calculate the area of an arbitrary polygon (and afterwards the area of a circular disk) can be regarded as a proof of the uniqueness (albeit one achieved only at the end of the study of the subject of area).

Finally, we note that when the students first encounter the problem of calculating the areas of polygons, they only know the formula for calculating the area of one polygon, namely the *rectangle*. This formula is first explained for rectangles whose sides can be expressed as finite decimals. And although the proof of the validity of the formula for *arbitrary* rectangles is not, as a rule, considered in high school, it is nevertheless assumed that the student "knows" the formula for the area of a rectangle in the general case.

Thus the starting point at which we have arrived, by studying the properties of area from the standpoint of contemporary mathematical ideas, corresponds rather closely to the level of knowledge of a high school student: *We know the properties* (α), (β), (γ^*), (δ) *of area, together with the formula for the area of a rectangle.* Starting from this point, we must find effective methods for *calculating* the area of an *arbitrary* polygon.

This is achieved by using a very simple method, called the *method of decomposition* and based on axioms (β) and (γ^*). To explain the method, we consider the figures F and H shown in Fig. 13 (all the line segments making up the figure F are congruent to one another, the angles are right angles, and the side of the square H is congruent to the segment ab). The dashed lines divide these figures into the same number of congruent pieces (corresponding congruent pieces are marked with the same number). Accordingly, we say that the figures F and H are *equidecomposable.* In other words, *the figures F and H are said to be equidecomposable if the figure F can be suitably decomposed into a finite number of pieces which can be reassembled to give the figure H* (by arranging the pieces differently, i.e., by considering figures congruent to the pieces).

Let us agree to express equidecomposability of two figures F and H by the symbol $\sim$, and congruence of two figures F_1 and H_1 by the

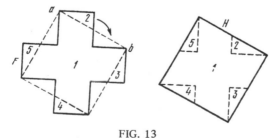

FIG. 13

symbol $\cong$. Thus, by definition, equidecomposability of the figures F and H (i.e., $F \sim H$) means that there exist figures $F_1, F_2, \ldots, F_k$ and $H_1, H_2, \ldots, H_k$, such that

$$F = F_1 + \ldots + F_h, \quad H = H_1 + \ldots + H_h;$$
$$F_1 \cong H_1, \ldots, F_k \cong H_k;$$

where all the figures $F, H, F_1, \ldots, F_k, H_1, \ldots, H_k$ are assumed to be measurable.

It is an immediate consequence of axioms (β) and (γ^*) that two equidecomposable figures *have the same area,* i.e., if $F \sim H$, then $s\,(F) = s\,(H)$. This is the basis for the *method of decomposition* (or *dissection*), which consists in calculating the area of a figure by trying to decompose it into a finite number of pieces in such a way that the pieces can be reassembled to form a *simpler* figure (whose area we already know). This method for calculating area was known as far back as the time of Euclid, who lived more than 2000 years ago.

We now recall some examples of the application of this method. A parallelogram F is equidecomposable with the rectangle H which has the same base and altitude (Fig. 14).[5] Therefore F and H have the same area, and hence, from a knowledge of the formula for the area of a rectangle, we find that the area of the parallelogram equals the product of the length of its base and the corresponding altitude.

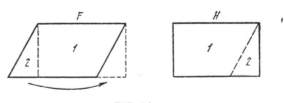

FIG. 14

[5] However, we note that this method (splitting off one triangle) is not always applicable (cf. Fig. 15).

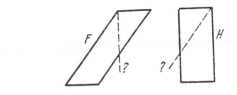

FIG. 15

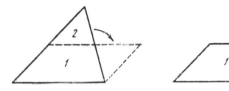

FIG. 16

A triangle is equidecomposable with the parallelogram which has the same base and half the altitude (Fig. 16). From this we deduce the formula for the area of a triangle. Finally, Fig. 17 shows a way of calculating the area of a trapezoid.

After deriving the formula for the area of a triangle, we can calculate the area of an arbitrary polygon. In fact, we need only decompose the polygon into triangles and then use axiom (β), i.e., add up the areas of the triangles. Note that any other way of decomposing the polygon into triangles leads to the same result, since both calculations give a *uniquely* determined number, namely the area s (F) of the polygon under consideration.[6]

[6] We observe that there is also another way of constructing a theory of area for polygons, in which the uniqueness and existence theorem is not used. Instead we *define* the area of an arbitrary polygon F by decomposing F into triangles and then setting the area $s(F)$ equal to the sum of the areas of the triangles making up F (the formula for the area of a triangle is regarded as an *axiom* in this approach). Here the greatest difficulties attend the proof of the fact that the area $s(F)$ is uniquely defined, i.e., that the sums of the areas of the triangles making up different decompositions of F are the same (cf. [45], pp. 195–204).

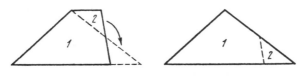

FIG. 17

There is still another way of calculating area, based on the use of axioms (β) and (γ*), known as the *method of complementation.* This method consists of adding congruent pieces to two figures in such a way that the resulting new figures are congruent. For example, to prove that the figures F and H shown in Fig. 13 have the same area, we could add four congruent triangles each to both the cross and the square (Fig. 18). Then, since the resulting figures are congruent, the original figures F and H have the same area.

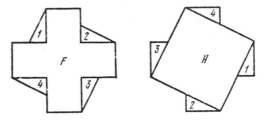

FIG. 18

Figure 19 shows that by adding one of a pair of congruent triangles to a parallelogram and the other to the rectangle with the same base and altitude, we get a pair of congruent trapezoids. Therefore the parallelogram and the rectangle have the same area. (We note that this method is always applicable, unlike the method of proof shown in Fig. 14.)

The method of complementation has already been used in Fig. 11 to prove property (γ*). In fact, the figures Q' and $Q_1 + Q_2$ were complemented by the addition of congruent triangles to make congruent squares. However, if the property (γ*) is taken as an

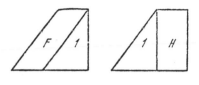

FIG. 19

axiom, then the method can be used to prove the Pythagorean theorem (a method of proof ascribed to the ancient Hindus). For comparison, we give a figure showing how the Pythagorean theorem is proved by the method of decomposition.

In general, we call two figures F and H *equicomplementable* if there exist figures $F_1, \ldots, F_k$ and $H_1, \ldots, H_k$ such that

$$F + F_1 + \ldots + F_k \cong H + H_1 + \ldots + H_h.$$

It is an immediate consequence of axioms (β) and (γ^*) that *equicomplementable figures have the same area.*

Thus, in the theory of area of polygons, axiom (α) (and the method of exhaustion) is used only *once,* in deriving the formula for

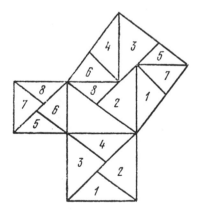

FIG. 20

the area of a rectangle. Once this formula has been established, the area of an *arbitrary* polygon can be calculated in an elementary fashion, by using the method of decomposition (or complementation), based only on axioms (β) and (γ*) (nothing is said about axiom (δ), since this axiom is implicit in the formula for the area of a rectangle).

The method of exhaustion is again needed to calculate the area of curvilinear figures, for example, the area of a circular disk and of its parts (see the footnote on p. 34). However, the general definition of area is not considered in high school, and the area of a disk is *defined* as the limit of the area of a regular polygon inscribed in the disk as the number of sides of the polygon is doubled without limit (for example, this is the approach adopted in the textbook [40]). As a result, the role of the "nonelementary" axiom (α) is obscured. And since, as a rule, the formula for the area of a rectangle is also given in high school without a full and accurate proof, the impression is created in the students that the theory of area is based only on axioms (β), (γ*), (δ), and that axiom (α) is unnecessary. As we saw in the preceding section, this point of view is incorrect, and axiom (α) is used in an essential way even in the theory of area of rectangles. However, if the formula for the area of a rectangle is regarded as an *axiom* (and this is essentially what is done in high school), calling it axiom (α**), say, then it can be said that the theory of area of polygons is constructed by using only the *three* axioms (α**), (β) and (γ*).

§6. The measurement of volume and Hilbert's third problem

The concept of volume is introduced by analogy with the concept of area. In the constructive definition of volume, we consider a *three-dimensional* mosaic, i.e., a decomposition of space into congruent cubes. For example, fixing a rectangular system of coordinates x, y and z, we can regard the kth mosaic as being the system of cubes into which space is decomposed by the planes $x = p/10^k$, $y = q/10^k$ and $z = r/10^k$, where p, q, r take all integral values. This allows us to consider the limits

$$\underline{v}(F) = \lim_{k \to \infty} \frac{a_k}{10^{3k}}, \quad \overline{v}(F) = \lim_{k \to \infty} \frac{b_k}{10^{3k}},$$

where a_k is the number of cubes of the kth mosaic contained in a solid figure F, and b_k is the number of cubes of the mosaic which have points in common with F. If the limits $\underline{v}(F)$ and $\overline{v}(F)$ coincide, then, by definition, F is *measurable* (or "cubable"), and in this case the number $v(F) = \underline{v}(F) = \overline{v}(F)$ is called the *volume* of the figure F.

Just as in the case of area, we can define volume axiomatically, with the axioms on which the concept of volume is based being completely analogous to the axioms of area. Here are the axioms in question:

(α) *The function* v *is nonnegative, i.e., the volume* $v(F)$ *of any measurable figure* F *is a nonnegative number.*

(β) *The function* v *is additive, i.e., if* F' *and* F'' *are measurable figures with no common interior points, then the figure* $F' \cup F''$ *is also measurable and* $v(F' \cup F'') = v(F') + v(F'')$.

(γ) *The function* v *is invariant under translations (parallel displacements), i.e., if* F *is a measurable figure and* F' *is the figure obtained by subjecting* F *to a translation, then the figure* F' *is also measurable and* $v(F') = v(F)$.

(δ) *The function* v *is normalized, i.e., the unit cube* Q *is a measurable figure and* $v(Q) = 1$.

By the unit cube in property (δ) we mean a fixed cube whose edges are of length 1.

If we take the constructive definition of volume (with the help of three-dimensional mosaics) as our starting point, then (α), (β), (γ) and (δ) are proved as theorems (the proofs are analogous to those given in §2).

A *polyhedron* can be defined either as a bounded closed set in space, whose boundary is the union of a finite number of plane polygons, or as the union of a finite number of *tetrahedra*, i.e., triangular pyramids. *Every polyhedron is a measurable figure.* The proof of this fact is analogous to the proof of Theorem 1. It is first proved that for every k we can construct a $1/10^k$-net in the boundary B of the polyhedron F, consisting of no more than $p \cdot 10^{2k} + q \cdot 10^k + m$ points (where p, q, m are numbers which are determined by the polyhedron F and do not depend on k). In fact, if a face Γ of the polyhedron F is placed on a square G of side a, then the vertices of the kth mosaic constructed in the plane of Γ form a $1/10^k$-net, and the square G contains no more than $(a \cdot 10^k + 1)^2$ points of this $1/10^k$-net. The rest of the proof is then virtually the same as that of Theorem 1 (with obvious changes).

The existence and uniqueness theorem, and also the necessary and sufficient condition for measurability of a solid figure, are proved by complete analogy to the proofs of Theorems 2, 3, 4, 5 in §2. All the facts presented in §3 generalize to the three-dimensional case, including the formula for the volume of a rectangular parallelepiped and property (γ^*) (the invariance of volume under arbitrary motions). The independence of axioms (α), (β), (γ^*), (δ) for volume is proved in exactly the same way as for area.

Axiom (α) serves as the basis for *nonelementary* methods of calculating volume (the method of exhaustion, integration), which are based on the use of a limiting process, while axioms (β) and (γ^*) serve as the basis for *elementary* methods (the methods of decomposition and complementation). It is necessary to use axiom (α) to derive the formula for the volume of a rectangular parallelepiped. However, if the lengths of the edges are rational, the formula for the volume of a rectangular parallelepiped can easily be derived without using axiom (α). Hence the formula is regarded as "known" when the subject of volume is studied in high school.

With the formula for the volume of a rectangular parallelepiped at our disposal, the question now arises of whether we can calculate the volume of an arbitrary polyhedron, using only the methods of decomposition and complementation, without resorting to the nonelementary method of exhaustion. Indeed, in the case of the area of an arbitrary polygon, we found that it was possible to do just this.

In certain cases, the method of decomposition (or complementation) can in fact be used to calculate the volume of a

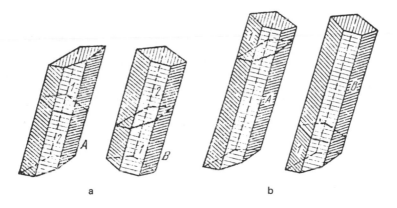

a   b

FIG. 21

polyhedron. For example, as shown in Fig. 21, an oblique prism is equidecomposable (and equicomplementable) with the right prism whose base is the perpendicular cross section of the oblique prism and whose lateral edge is of the same length as that of the oblique prism. The volume of a right prism can in turn be calculated by using the method of decomposition (or complementation). In fact, the right prism whose base is a parallelogram is equidecomposable (and equicomplementable) with a rectangular parallelepiped, as shown in Fig. 22 (cf. Fig. 14). Moreover, a right prism with a triangular base is equidecomposable with a prism whose base is a parallelogram, as

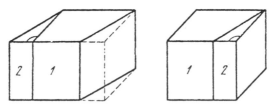

FIG. 22

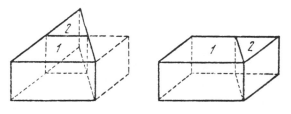

FIG. 23

shown in Fig. 23 (cf. Fig. 16). Finally, an arbitrary right prism can be decomposed into prisms with triangular bases (Fig. 24). Thus the formula $v = sh$ (where v is the volume of a right prism with altitude h and base of area s) can be proved without resorting to the "nonelementary" axiom (α).

On the other hand, in calculating the volume of a *pyramid*, the textbook literature uses the method of exhaustion: A series of "step-shaped objects" is considered (see Fig. 1), and then the limit is taken as the number of steps is increased without limit (the "devil's staircase").

Let us now recall briefly how the formula for the volume of a tetrahedron is usually derived. We take a tetrahedron *abcd*, and consider the oblique prism with triangular base *abc* and lateral edge *ad* (Fig. 25). This prism can be decomposed into three tetrahedra F_1, F_2, F_3 (Fig. 26), of which the first two, and also the last two, have congruent bases and equal altitudes. Thus, to prove that the volume of the pyramid *abcd* is *one third* the volume of the prism in

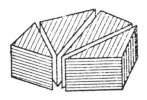

FIG. 24

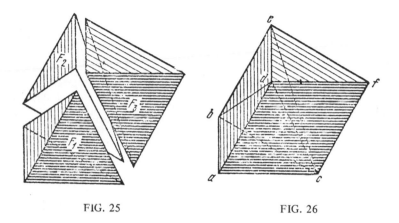

FIG. 25   FIG. 26

question (i.e., to derive the formula for the volume of a pyramid), "it only remains" to prove that *two triangular pyramids with congruent bases and equal altitudes have equal volume.* But this is proved by using the method of exhaustion (or integration).

Thus, as compared with the theory of area of polygons, the theory of volume of polyhedrons contains an "extra" application of axiom (α). Although the formula for the area of a triangle has a completely elementary derivation, the spatial analogue, namely the formula for the volume of a tetrahedron, relies on the use of axiom (α). Is this essential? Or is this way of doing things due only to the fact that mathematicians have not yet been "lucky enough" to find a simple proof of the formula for the volume of a pyramid, by using the method of decomposition or complementation? In other words, are any two triangular pyramids with congruent bases and equal altitudes equidecomposable (or equicomplementable)? This is precisely *Hilbert's third problem,* which we now cite in its original formulation ([33], p. 449; for exact references to the cited papers by Dehn, see [12]):

"In two letters to Gerling, Gauss* expresses his regret that certain theorems of solid geometry depend upon the method of exhaustion,

*Werke, vol. 8, pp. 241 and 244.

i.e., in modern phraseology, upon the axiom of continuity (or upon the axiom of Archimedes). Gauss mentions in particular the theorem of Euclid, that triangular pyramids of equal altitudes are to each other as their bases. Now the analogous problem in the plane has been solved.* Gerling also succeeded in proving the equality of volume of symmetrical polyhedra by dividing them into congruent parts. Nevertheless, it seems to me probable that a general proof of this kind for the theorem of Euclid just mentioned is impossible, and it should be our task to give a rigorous proof of its impossibility. This would be obtained as soon as we succeeded in *exhibiting two tetrahedra of equal bases and equal altitudes which can in no way be split up into congruent tetrahedra, and which cannot be combined with congruent tetrahedra to form two polyhedra which themselves could be split up into congruent tetrahedra.†*"

*Cf., beside earlier literature, Hilbert, Grundlagen der Geometrie, Leipzig, 1899, ch. 4. [Translation by Townsend, Chicago, 1902.]

† Since this was written Herr Dehn has succeeded in proving this impossibility. See his note: "Ueber raumgleiche Polyeder," in *Nachrichten d. K. Gesellsch. d. Wiss. zu Göttingen,* 1900, and a paper soon to appear in the *Math. Annalen* [vol. 55, pp. 465–478].

CHAPTER 2

EQUIDECOMPOSABILITY
OF POLYGONS

§7. The Bolyai–Gerwien theorem

The reader doubtlessly noticed Hilbert's sentence that says "Now the analogous problem in the plane has been solved." Of course, here Hilbert is not alluding to the simple fact that the formula for the area of a triangle (unlike the formula for the volume of a pyramid) can be derived by using the method of decomposition (or complementation), a fact already familiar to Euclid. The meaning of Hilbert's sentence is revealed by the accompanying footnote, in which he mentions the "earlier literature" and the fourth chapter of his book "Foundations of Geometry" [32]. The material in Chapter IV of reference [32] shows that Hilbert had in mind the complete solution, found in the last century, of the problem of the interrelation between the concepts of *equality of area, equidecomposability* and *equicomplementability* for polygons.

And what does this interrelation consist of? As noted in §5, *any two equidecomposable polygons have the same area*, i.e., if $F \sim H$, then $s(F) = s(H)$. It is natural to ask the converse question: Is every pair of polygons with the same area equidecomposable, i.e.,

49

does $s(F) = s(H)$ always imply $F \sim H$? It was found independently by the Hungarian mathematician F. Bolyai[7] (in 1832) and by the German officer and mathematical amateur P. Gerwien (in 1833) that this question has an affirmative answer. More exactly, the *Bolyai-Gerwien theorem* asserts that *two polygons have the same area if and only if they are equidecomposable.* It is also an easy consequence of this theorem that two polygons have the same area if and only if they are equicomplementable. Thus *equality of area, equidecomposability* and *equicomplementability* are equivalent properties of polygons in Euclidean planimetry. The present section is devoted to an exposition of these results.

Lemma 1. *If $A \sim B$ and $B \sim C$, then $A \sim C$.*

In other words, if each of the figures A and C is equidecomposable with B, then A and C are also equidecomposable. In fact, suppose we draw the lines decomposing figure B into pieces which can be rearranged to give figure A (the solid lines in Fig. 27a), and then draw the lines decomposing figure B into pieces which can be rearranged to give figure C (the solid lines in Fig. 27b). Then these two sets of lines together decompose B into smaller pieces which can be rearranged to give both A and C. Thus $A \sim C$.

This proof can be formalized as follows. Let

$$A = A_1 + \ldots + A_k, \quad B = B_1 + \ldots + B_k =$$
$$= B_1' + \ldots + B_l', \quad C = C_1 + \ldots + C_l;$$

and suppose further that the formulas

$$A_1 = f_1(B_1), \ldots, A_k = f_k(B_k), \quad C_1 = g_1(B_1'), \ldots$$
$$\ldots, C_l = g_l(B_l')$$

hold, where $f_1, \ldots, f_k, g_1, \ldots, g_l$ are certain motions (so that $A_1 \cong B_1$, $C_1 \cong B_1'$, and so on). These formulas just mean that each of the figures A and C is equidecomposable with B. Let $F_{ij} = B_i \cap B_j'$, where $i = 1, \ldots, k; j = 1, \ldots, l$ (note that some

---

[7] The father of the famous J. Bolyai, who arrived at the ideas of non-Euclidean geometry independently of N. I. Lobachevskiy.

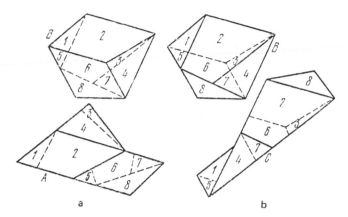

FIG. 27

of the figures F_{ij} may be empty). It is clear that no two figures F_{ij} have common interior points; for example, if $i^* \neq i$, then $F_{i^*j^*} \subset B_{i^*}$, $F_{ij} \subset B_i$, and since B_{i^*} and B_i have no common interior points, the same is true of $F_{i^*j^*}$ and F_{ij}. Moreover, we have

$$\bigcup_{j=1}^{l} F_{ij} = \bigcup_{j=1}^{l} (B_i \cap B_j') = B_i \cap (\bigcup_{j=1}^{l} B_j') = B_i \cap B = B_i.$$

Therefore

$$A = \bigcup_{i=1}^{k} A_i = \bigcup_{i=1}^{k} f_i(B_i) = \bigcup_{i=1}^{k} f_i(\bigcup_{j=1}^{l} F_{ij}) = \bigcup_{i=1}^{k} \bigcup_{j=1}^{l} f_i(F_{ij}),$$

and similarly

$$C = \bigcup_{i=1}^{k} \bigcup_{j=1}^{l} g_j(F_{ij}).$$

Finally, we need only note that $f_i(F_{ij}) \cong g_j(F_{ij})$ (the motion $g_j \circ f_i^{-1}$ carries the first of these figures into the second), and hence $A \sim C$.

From the proof it is apparent that here A, B, C, and also A_i, B_i, B_j', C_j can be arbitrary measurable figures; the figure F_{ij} will then be measurable, by Theorem 4. In the case where B_i and B_j' are *polygons*, the intersection $F_{ij} = B_i \cap B_j'$ is also a polygon. In this case we can talk about the "lines" decomposing the figure B into pieces $B_1, \ldots, B_k$ (or $B_1', \ldots, B_l'$). Henceforth we will need only the case where all the figures under consideration are polygons.

Lemma 2. *Every triangle is equidecomposable with some rectangle.*

In fact, let ab (say) be the largest side of the given triangle abc, and draw the altitude from c to ab (Fig. 28). The point d lies *between* a and b (otherwise the side ab would not be the largest; see Fig. 29). Through the midpoint of cd we draw the line $mn \parallel ab$ and drop perpendiculars ae and bf to the line mn. This gives a rectangle $aefb$ which is equidecomposable with the triangle abc. In fact, the triangles marked 1 in Fig. 28 are congruent, and so are those marked 2. But each of the figures abc and $aefb$ consists of the shaded trapezoid and a pair of triangles marked 1 and 2.

Lemma 3. *Any two rectangles with the same area are equidecomposable.*

In fact, we arrange the rectangles of equal area (the rectangles $oabc$ and $omnp$ in Fig. 30) in such a way that they have a common right angle. Let the lengths of the segments oc, oa, op, om be denoted by l_1, h_1, l_2, h_2, respectively. Then $l_1 h_1 = l_2 h_2$ (since the rectangles have the same area), i.e., $\dfrac{l_1}{h_2} = \dfrac{l_2}{h_1}$, which implies

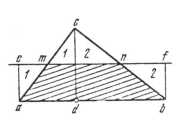

FIG. 28

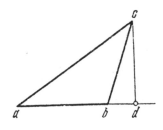

FIG. 29

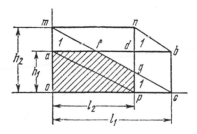

FIG. 30

$ap \parallel mc$. Moreover, $\dfrac{l_1 - l_2}{h_2 - h_1} = \dfrac{l_2}{h_1}$, i.e., the triangles oap and dnp are similar, and hence $nb \parallel ap$. Thus all three lines ap, mc, nb are parallel. If the segment mc *intersects* the rectangle $oadp$, then the equidecomposability of the rectangles $oabc$ and $omnp$ is obvious: Each of them consists of the shaded pentagon in Fig. 30, one of the two congruent triangles marked 1, and one of the two congruent triangles gmn and cfb.

We now consider the case where the segment mc does not intersect the rectangle $oadp$ (Fig. 31). Since the segments mn, ad and fb are congruent, in this case the sum of the lengths of the segments ad and fb is smaller than the length of the segment ab, i.e., $2l_2 < l_1$ (whereas $2l_2 \geqslant l_1$ in the case shown in Fig. 30). Let e be

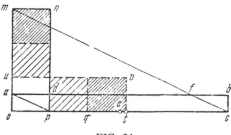

FIG. 31

the midpoint of the segment oc, and let k be the smallest natural number such that after laying off k segments $op, pq \ldots$ congruent to op along the line oc we first get a point t lying outside the segment oe, so that the point t lies inside the segment ec. (In Fig. 31 we have $k = 3$.) Next, drawing lines parallel to op, we subdivide the rectangle $omnp$ into k congruent parts and lay these parts along the segment ot. This gives a rectangle $ouvt$ (obviously equidecomposable with $omnp$) whose base ot is of length l' satisfying the condition $2l' > l_1$. But the rectangles $oabc$ and $ouvt$ are equidecomposable, by the first part of the proof. Therefore, the rectangles $oabc$ and $omnp$ are equidecomposable, by Lemma 1.

Theorem 10 (Bolyai–Gerwien theorem). *Any two polygons with the same area are equidecomposable.*

PROOF. Every polygon F can be decomposed into a finite number of triangles, and each of these triangles is equidecomposable with some rectangle, by Lemma 2. Thus $F \sim P_1 + \ldots + P_k$, where $P_1, \ldots, P_k$ are rectangles. Choosing an arbitrary line segment $a_0 b_0$, we erect perpendiculars $a_0 c$ and $b_0 d$ at its ends (Fig. 32). Then we draw segments $a_1 b_1, a_2 b_2, \ldots, a_k b_k$, parallel to $a_0 b_0$, in such a way that the rectangle $a_{i-1} a_i b_i b_{i-1}$ (which we denote by H_i) has the *same area* as P_i, $i = 1, \ldots, k$. Since $s(P_i) = s(H_i)$, we have $P_i \sim H_i$, by Lemma 3. Therefore, $P_1 + \ldots + P_k \sim H_1 + \ldots + H_k$, so that, by Lemma 1, F is equidecomposable with $H_1 + \ldots + H_k$, i.e., with the rectangle $a_0 a_k b_k b_0$.

Thus every polygon is equidecomposable with some rectangle. If now F and G are two polygons such that $s(F) = s(G)$, then we can find rectangles P and Q, equidecomposable with F and G, such that $F \sim P$, $G \sim Q$. The rectangles P and Q have the same area. Therefore, $P \sim Q$, by Lemma 3. Thus $F \sim P, P \sim Q, Q \sim G$, and hence, by Lemma 1, the polygons F and G are equidecomposable.

Theorem 11. *If two figures are equidecomposable, then they are equicomplementable.*

PROOF. Given $A \sim B$, we set $C = A \cap B$. Let M and N be the figures such that $A = C + M$, $B = C + N$. Moreover, let A^* and B^* be figures such that $A \cong A^*$, $B \cong B^*$, while $A \cup B$, A^* and B^* are pairwise disjoint. Since $A \sim B$, we have $A^* \sim B^*$,

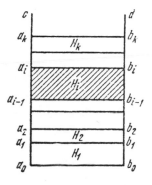

FIG. 32

i.e., there exist figures $P_1, \ldots, P_k, Q_1, \ldots, Q_k$, such that

$$A^* = P_1 + \ldots + P_k, \quad B^* = Q_1 + \ldots + Q_k,$$
$$P_1 \cong Q_1, \ldots, P_k \cong Q_k.$$

If f and g are motions such that $A^* = f(A)$, $B^* = g(B)$, then

$$(A \cup B) + A^* + B^* = (A + N) + f(A) + Q_1 + \ldots$$
$$\ldots + Q_k = A + N + Q_1 + \ldots + Q_k + f(C) + f(M),$$
$$(A \cup B) + A^* + B^* = (B + M) + P_1 + \ldots + P_k +$$
$$+ g(B) = B + M + P_1 + \ldots + P_k + g(C) + g(N).$$

Therefore

$$A + Q_1 + \ldots + Q_k + f(M) + N + f(C) =$$
$$= B + P_1 + \ldots + P_k + M + g(N) + g(C),$$

from which it is clear that A and B are equicomplementable.

Theorems 10 and 11, which we have just proved, establish the equivalence of the concepts of "equality of area," "equidecomposability" and "equicomplementability" for polygons. In fact, if two polygons have the same area, then they are equidecomposable (Theorem 10) and hence equicomplementable (Theorem 11). The converse is an immediate consequence of axioms (β) and (γ^*). We also note that we could have taken equicomplementability to have a more general meaning, in which the complementation in question uses congruent figures to give *equidecomposable* figures rather than congruent figures (it is clear from Hilbert's formulation of the problem that he mean equicomplementability in just this sense). But it follows from Theorem 11 that this "generalized" equicomplementability is equivalent to ordinary equicomplementability.

§8. Equidecomposability and equicomplementability in non-Archimedean and non-Euclidean geometries

In Lemma 3, which is the key part of the proof of the Bolyai–Gerwien theorem, it is assumed that in laying off congruent segments *op, pq,* ... along *oe,* we eventually obtain (after k applications) a point lying *outside* the segment *oe.* In Chapter IV of reference [32] Hilbert shows, by an elegant example, that the use of the Archimedean axiom (or some other axiom of continuity) cannot be avoided in proving the theorem. In fact, in so-called *non-Archimedean* geometries equality of area and equicomplementability remain equivalent concepts (for polygons), whereas equidecomposability is *no longer* equivalent to these concepts.

Of course, in proving the Bolyai–Gerwien theorem, other axioms of geometry were used as well, for example, the axiom of parallelism (since we considered *rectangles*). However, unlike the case of the Archimedean axiom, application of the axiom of parallelism can be avoided: The Bolyai–Gerwien theorem remains valid in the non-Euclidean geometries of Lobachevsky and Riemann. The present section is devoted to an exposition of these results.

We will consider power series of the form

$$x = t^k (a_0 + a_1 t + a_2 t^2 + \ldots), \qquad a_0 \neq 0, \qquad (7)$$

where k is an *integer* and the series in parentheses converges for some nonzero value of t (and hence converges absolutely and uniformly on a closed interval $[-\varepsilon, \varepsilon]$, where ε is some positive number). Let Ω denote the set consisting of all series of the indicated type, together with the null series 0. The sum, difference, product and quotient of two series of the form (7), calculated by the usual rules for dealing with power series, are again series of the form (7). Thus the set Ω is a *field*.

Let us agree to write $x > 0$ if $a_0 > 0$ in (7), and $x > y$ if $x - y > 0$. Given arbitrary elements x, $y \in \Omega$, one and only one of the relations $x > y, x = y, y > x$ holds. It is clear that the sum and product of positive elements of the field Ω are positive, from which all the properties of inequalities follow. Thus Ω is an ordered field. This field is *non-Archimedean,* i.e., we can find positive elements x, $y \in \Omega$ such that the inequality $nx > y$ does not hold for any positive integer n whatsoever. In fact, we need only choose positive elements x (see (7)) and

$$y = t^l (b_0 + b_1 t + b_2 t^2 + \ldots), \qquad b_0 \neq 0,$$

such that $l < k$. We also note that the element $\sqrt{x}$ is defined for every nonnegative element $x \in \Omega$ for which k (see (7)) is even, i.e., for such x there exists a unique nonnegative element z satisfying the condition $z^2 = x$. In particular, the element $\sqrt{x^2 + y^2}$ is defined for arbitrary x, $y \in \Omega$.

Using the field Ω, we can construct a *non-Archimedean geometry* (we confine ourselves to the case of planimetry). In fact, let us agree to call every pair (x_0, y_0) with x_0, $y_0 \in \Omega$ a *point*. A *line* is the set of all points (x_0, y_0) whose coordinates satisfy an equation of the form $ax + by + c = 0$ (where a, b, $c \in \Omega$ and $a^2 + b^2 \neq 0$). By a *motion* we mean a transformation carrying every point (x, y) into a point (x', y') in accordance with the formulas

$$x' = ax + by + p,$$
$$y' = cx + dy + q,$$

where a, b, c, d, p, q are elements of the field Ω, satisfying the conditions

$$a^2 + c^2 = 1, \quad b^2 + d^2 = 1, \quad ab + cd = 0$$

(these conditions mean that the matrix of the transformation is orthogonal, i.e., that it preserves the scalar product of vectors).

As usual, two figures are said to be *congruent* if one of them is carried into the other by a motion. The *distance* between the points $a = (x_1, y_1)$ and $b = (x_2, y_2)$ is defined by the formula

$$d(a, b) = \sqrt{(x_2 - x_1)^2 + (y_2 - y_1)^2}.$$

This distance (which is an element of Ω) is invariant under motions. In fact, the square of the distance is the scalar product of the vector $\overrightarrow{ab}$ with itself, and scalar products are invariant under motions. In this geometry we can also talk about parallel lines (the vectors directed along such lines are proportional), about perpendicular lines (the vectors directed along such lines are orthogonal, i.e., their scalar product equals zero), about triangles, parallelograms and rectangles, and about the lengths of their sides and altitudes (these lengths are positive elements of the field Ω).

Finally, we arrive at the study of areas. We *assume* that the area of a rectangle is given by the formula $s = ab$, where a and b are the lengths of its sides, and s is its area. From this (cf. Figs. 19, 16) we deduce the usual formulas for the area of a parallelogram and a triangle. We can then decompose any polygon into triangles, which allows us to calculate its area. As a result, we obtain a function $s(F)$, the area, defined on the set of all polygons. It can be proved that this function satisfies axioms (α), (β), (γ^*), (δ). However, it is not the *unique* function satisfying these axioms. In fact, given any element $x \in \Omega$ (see (7)), let

$$f(x) = t^{2k}(a_0 + a_1 t^2 + a_2 t^4 + \ldots).$$

Clearly $f(x) \in \Omega$. There is no difficulty in verifying that the function $f(x)$ is *additive*, i.e., satisfies the condition (3), p. 26.

Moreover, it is clear that $f(x) > 0$ whenever $x > 0$. Thus, unlike the Cauchy-Hamel functions, this (nonlinear) additive function is *positive* on all positive elements of the field Ω. We also note that $f(1) = 1$ (where 1 is the unit of the field Ω, i.e., the series $t^0 (1 + 0 \cdot t + 0 \cdot t^2 + \ldots)$). Finally, we set

$$s^* (F) = f(s(F)).$$

Then, as can be verified at once, the function $s^* (F)$ also satisfies axioms (α), (β), (γ^*), (δ), thereby proving the nonuniqueness of the function $s(F)$. However, if instead of (α) and (δ), we include the formula for the area of a rectangle as one of the *axioms*, then axioms (β) and (γ^*) together with this new axiom, *uniquely* determine the area (on the set of polygons).

We now give Hilbert's example, showing the *nonequivalence* of the concepts of equality of area and equidecomposability in non-Archimedean geometry. Suppose that on a ray we lay off two segments ab and ad whose lengths e and l are such that $ne > l$ does not hold for any positive integer whatsoever. Moreover, let ac and dq be segments of length e perpendicular to the line ab (Fig. 33). Then the two triangles abc and abq of equal area (they have a common base and equal altitudes) *are not* equidecomposable. In fact, suppose each of them can be decomposed into a finite number of triangles. Then each of the subtriangles making up the decomposition of the triangle abc has a perimeter less than that of the triangle abc, i.e., a perimeter certainly less than $4e$ (since the length of the side bc is less than the sum of the lengths of the other two sides of the triangle abc). Hence the sum of the perimeters of these subtriangles is less than $4ke$, where k is the number of subtriangles making up the

FIG. 33

decomposition of the triangle *abc*. But $4ke < l$, and hence $4ke$ is less than the length of the side *aq*. Hence the sum of the perimeters of the subtriangles making up the decomposition of the triangles *abq* is *greater* than $4ke$. But then the subtriangles making up the decompositions of the triangles *abc* and *abq* cannot be pairwise congruent.

Thus, in non-Archimedean geometry, triangles of equal area may not be equidecomposable. At the same time, equality of area and equicomplementability are still equivalent concepts, as we will see in a moment. In proving this, we introduce the notation $A \overset{c}{\sim} B$, meaning that the polygons A and B are equicomplementable.

Lemma 1$_c$. *If $A \overset{c}{\sim} B$ and $B \overset{c}{\sim} C$, then $A \overset{c}{\sim} C$.*

In fact, suppose that

$$A + f_1(M_1) + \ldots + f_h(M_h) = f^*(B + M_1 + \ldots + M_h),$$

$$C + g_1(N_1) + \ldots + g_l(N_l) = g^*(B + N_1 + \ldots + N_l),$$

where f_i, g_j, f^*, g^* are motions, and let

$$P_{ij} = M_i \cap N_j, \quad M_i^* = \overline{M_i \backslash (N_1 \cup \ldots \cup N_l)},$$

$$N_j^* = \overline{N_j \backslash (M_1 \cup \ldots \cup M_h)}.$$

Then

$$A + \sum_i f_i(M_i^*) + \sum_{i,j} f_i(P_{ij}) + \sum_j f^*(N_j^*) =$$

$$= A + \sum_i f_i\left(M_i^* + \sum_j P_{ij}\right) + \sum_j f^*(N_j^*) =$$

$$= A + \sum_i f_i(M_i) + \sum_j f^*(N_j^*) =$$

$$= f^*\left(B + \sum_i M_i\right) + \sum_j f^*(N_j^*) = f^*\left(B + \sum_i M_i + \sum_j N_j^*\right) =$$

$$= f^*\left(B \cup \sum_i M_i \cup \sum_j N_j\right).$$

and similarly

$$C + \sum_j g_j(N_j^*) + \sum_{i,j} g_j(P_{ij}) + \sum_i g^*(M_i^*) =$$
$$= g^* \left(B \cup \sum_i M_i \cup \sum_j N_j \right).$$

Thus the polygons

$$A + \sum_i f_i(M_i^*) + \sum_j f^*(N_j^*) + \sum_{i,j} f_i(P_{ij})$$

and

$$C + \sum_i g^*(M_i^*) + \sum_j g_j(N_j^*) + \sum_{i,j} g_j(P_{ij})$$

are congruent, and hence equicomplementable, by Theorem 11. But this immediately implies $A \overset{c}{\sim} C$.

Lemma 2$_c$. *Every triangle is equicomplementable with some rectangle.*

This follows from Lemma 2 and Theorem 11 (whose proofs do not make use of the Archimedean axiom).

Lemma 3$_c$. *Any two rectangles with the same area are equicomplementable.*

We cannot deduce Lemma 3$_c$ from Lemma 3 and Theorem 11, since Lemma 3 is based on the Archimedean axiom. Figure 34 illustrates a proof making no use of this axiom: The fact that the rectangles F and G have the same area implies that the triangles *bmp*

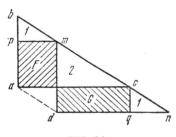

FIG. 34

and *cnq* are congruent. Therefore, if we add the triangles 1 and 2 to each of the rectangles F and G, we get a pair of congruent figures, namely the triangles *abc* and *dmn*.

Lemma 4$_c$. *If* $M_1 \overset{c}{\sim} N_1, \ldots, M_k \overset{c}{\sim} N_k$, *where* $M_1, \ldots M_k$ *have pairwise disjoint interiors and the same is true of* $N_1, \ldots, N_k$, *then* $M_1 + \ldots + M_k \overset{c}{\sim} N_1 + \ldots + N_k$.

We note that the analogous lemma for the case of equidecomposability was actually used in the proof of the Bolyai–Gerwien theorem, but was not explicitly formulated because of its obvious nature.

We give the proof for $k = 2$ (the rest follows by an obvious induction). Choose a motion f such that the interiors of all the polygons needed to establish the relation $M_1 \overset{c}{\sim} N_1$ are pairwise disjoint from the interiors of all the polygons needed to establish the relation $f(M_2) \overset{c}{\sim} f(N_2)$ (i.e., the motion f carries the pair of polygons M_2, N_2 "far enough away" from M_1, N_1). It is clear that

$$M_1 + M_2 \overset{c}{\sim} M_1 + f(M_2), \quad N_1 + N_2 \overset{c}{\sim} N_1 + f(N_2); \quad (8)$$

for example, by adding the congruent figures $f(M_2)$ and M_2 to $M_1 + M_2$ and $M_1 + f(M_2)$, we get the polygon $M_1 \cup M_2 \cup f(M_2)$. Moreover, it follows from the relations $M_1 \overset{c}{\sim} N_1$, $f(M_2) \overset{c}{\sim} f(N_2)$ that

$$M_1 + f(M_2) \overset{c}{\sim} N_1 + f(M_2), \quad N_1 + f(M_2) \overset{c}{\sim} N_1 + f(N_2). \quad (9)$$

But (8) and (9) imply $M_1 + M_2 \overset{c}{\sim} N_1 + N_2$ because of Lemma 1_c.

Theorem 10$_c$. *Any two polygons with the same area are equicomplementable.*

The proof of Theorem 10_c is virtually the same as that of Theorem 10. We need only replace the word "equidecomposable" everywhere by the word "equicomplementable" and use Lemmas 1_c–4_c.

Finally, we turn to the problem of clarifying the role of the postulate of parallels. The study of area in hyperbolic and elliptic geometry is based on four analogous axioms (where axiom (δ) must be modified, since there are no squares in these geometries). Just as in Euclidean geometry, axioms (α), (β), (γ^*) define a function $s(F)$ (area) to within a positive factor. Therefore an axiom similar to (δ) is needed to fix this factor uniquely. In hyperbolic (Lobachevskian) geometry, the appropriate axiom is

(δ_h) *The infinite "triangle" bounded by three mutually parallel lines* (Fig. 35) *is of area* π,

while in elliptic (Riemannian) geometry, it is

(δ_e) *The whole plane is of area* 2π.

To prove the existence of an area function, we use the concept of the *defect* (synonymously, "angular defect") of a polygon. If F is a polygon bounded by a simple closed polygonal curve with n vertices, then its defect $D(F)$ is defined as the number $\pi(n-2) - \Sigma$, where Σ is the sum of the interior angles of F (for polygons of a more complex form, the definition is modified in an obvious way). The defect of an arbitrary polygon is positive in hyperbolic geometry, and negative in elliptic geometry (the number $-D(F)$ is called the angular *excess* of the polygon F). It can be immediately verified that the function $s(F) = |D(F)|$ satisfies axioms (α), (β), (γ^*) and a simple passage to the limit shows that the function $s(F)$ satisfies

FIG. 35

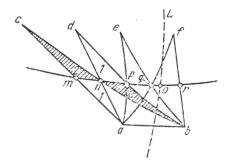

FIG. 36

axiom (δ_h) in Lobachevskian geometry (every "angle" of the infinite "triangle" in Fig. 35 equals zero) and axiom (δ_e) in Riemannian geometry. This proves the existence theorem. The proof of the uniqueness theorem will be considered later.[8]

Lemma 5. *Any two triangles with the same base and equal defects are equidecomposable.*

In fact, let mn be the midline of the triangle abc (Fig. 36). Under reflection in the point n, the point a goes into a point d, while the point m goes into a point p of the line mn. It is clear that n is the midpoint of the segment ad, while p is the midpoint of the segment bd (since the points b, d, p are symmetric to the points c, a, m). Thus np is the midline of the triangle abd, i.e., the midlines mn and np lie on the same line and are congruent. From the congruence of the triangles in Fig. 36 marked with the same number 1, it follows that the triangles abc and abd are equidecomposable (and hence have the same defect).

We can repeat this construction, carrying out a reflection in the point p (thereby obtaining the triangle abe, which is equidecomposable with the triangle abc and has a midline congruent to

---

[8] In proving the uniqueness theorem with the help of Lemmas 5–6 and Theorem 12, we follow the ideas of Uspensky's presentation of the theory of area, to be found in his book [55].

mn), and so on. By the Archimedean axiom, after a finite number of steps we get a triangle (*abf* in Fig. 36) with base *ab*, which is equidecomposable with the triangle *abc* and whose midline has a point *o* in common with the perpendicular *L* to the segment *ab* at its midpoint.

Let *abf* be the triangle so obtained (Fig. 37), and drop perpendiculars *au*, *bv*, *fw* onto its midline *qr*. The segments *uq* and *wq* are symmetric with respect to the point *q* and hence congruent; in exactly the same way, *vr* and *wr* are congruent. Under a homothetic transformation with center *w* and ratio 1/2, the points *u*, *v* go into *q*, *r*, and therefore *qr* is *half* the length of *uv*. In other words, the segment *uv* is of length 2*l*, where *l* is the length of the segment *qr*. Because of the symmetry, the segments *au* and *fw* are congruent, and so are *fw* and *bv*. Therefore *au* and *bv* are congruent, and hence the quadrilateral *abvu* is symmetric with respect to the line *L*. It follows that the distance between the point *o* and each of the points *u*, *v* equals *l*. But the distance between *o* and the points *q*, *r* does *not exceed l* (since the point *o* belongs to the segment *qr* of length *l*). Thus both points *q* and *r* belong to the segment *uv*. Moreover, by considerations of symmetry, the triangles in Fig. 37 marked with the same numbers are congruent. Therefore the triangle *abf*, and hence the triangle *abc* as well, is equidecomposable with the quadrilateral *abvu*.

We now note that the quadrilateral *abvu* is *uniquely* determined by the original triangle *abc*. In fact, the side *ab* is given, and

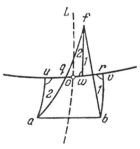

FIG. 37

moreover each of the angles a and b of the quadrilateral $abvu$ is equal to $\frac{1}{2}(\pi - \Delta)$, where Δ is the defect of the triangle abc (since abc and $abvu$ are equidecomposable and therefore have the same defect). Hence the position of the lines au and bv is also known. Finally, the line uv is uniquely determined as the common perpendicular of the lines au and bv.

Thus if abc' is any triangle with the same base ab and the same defect Δ as abc, then abc' is equidecomposable with the same quadrilateral $abvu$, and hence the triangles abc and abc' are equidecomposable.

Lemma 6. *Any two triangles with the same defect are equidecomposable.*

In fact, let abc and mnp be two triangles with defect Δ, and, to be explicit, let the segment ab be the largest (or one of the largest) of the six sides of these triangles. If any of the sides of the triangle mnp, say mn, is congruent to ab, then the triangles abc, mnp have congruent bases ab, mn and equal defects, and hence are equidecomposable, by the preceding lemma.

Now suppose that every side of the triangle mnp is *less* than ab. On the ray mp we lay off a segment mq congruent to ab (Fig. 38), and inside the angle pmk, adjacent to the angle pmn, we draw an arc of the circle with center m going through the point q. Let the point x traverse this arc from the point q to the point l lying on the ray

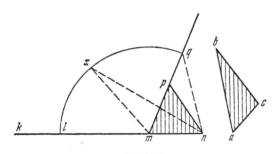

FIG. 38

mk. Then the defect of the resulting triangle *mnx* depends continuously on the position of the point *x*. When the point *x* is near *l*, the defect of the triangle *mnx* is near zero (since the angle *xmn* is near π and the other two angles are near zero). If $x = q$, however, then the defect of the triangle *mnx* (equal to the sum of the defects of the triangles *mnp* and *npq*) will exceed the defect of the triangle *mnp* in absolute value. Hence there is a position of the point *x* on the arc in question where the defect of the triangle *mnx* equals Δ.

Since the triangles *abc*, *mnx* have congruent sides *ab*, *mx*, and the same defect, they are equidecomposable (Lemma 5). The triangles *mnx* and *mnp* are equidecomposable for the same reason (they have a common side *mn* and the same defect). Hence the triangles *abc* and *mnp* are equidecomposable.

Theorem 12. *Any two polygons with the same defect are equidecomposable, in both hyperbolic and elliptic geometry.*

PROOF. Given any two polygons F and G with the same defect, we decompose both of them into triangles:

$$F \rightarrow T_1 + \ldots + T_m, \quad G = T_1' + \ldots + T_n'.$$

To be explicit, suppose that of all the triangles T_i and T_j', the triangle T_m has the defect of smallest absolute value, so that, in particular,

$$|D\,(T_m)| \leqslant |D\,(T_n')|. \tag{10}$$

If *equality* holds here, then $T_m \sim T_n'$, by Lemma 6, and we need only prove the equidecomposability of the polygons

$$F^* = T_1 + \ldots + T_{m-1} \text{ and } G^* = T_1' + \ldots + T_{n-1}',$$

which are divided into *fewer* than $m + n$ triangles. On the other hand, if equality does not hold in (10), i.e., if $|D\,(T_m)| < |D\,(T_n')|$, then we can draw a line through a vertex of the triangle T_n', dividing it into two triangles T_n^* and T_m'', the first of which has a defect *equal* to $D\,(T_m)$. Hence, by Lemma 6, the triangles T_m and T_n^* are equidecomposable, and we need only prove the equidecomposability of the polygons

$$F^* = T_1 + \ldots + T_{m-1} \text{ and } G^{**} = T_1' + \ldots + T_{n-1}' + T_n',$$

which are again divided into *fewer* than $m + n$ triangles.

Using this method of decreasing the number of triangles, we can now complete the proof, with the help of an obvious induction.

Corollary (Uniqueness theorem). *In hyperbolic (or elliptic) geometry there exists a unique function $s\ (F)$ which is defined on the set of all polygons and satisfies axioms (α), (β), (γ^*), (δ_h) (or (α), (β), (γ^*), $(\delta_e))$, in the elliptic case). This function is $s\ (F) = |\ D\ (F)\ |$.*

We give the proof for the hyperbolic case. Suppose there exists a function $s^*\ (F)$ which is different from $D\ (F)$ and satisfies axioms (α), (β), (γ^*), (δ_h). Choose a polygon F_0 for which $D\ (F_0) \neq s^*\ (F_0)$, and suppose, to be explicit, that $D\ (F_0) > s^*(F_0)$. If the polygon F satisfies the condition $D\,(F) = \frac{1}{q}\,D(F_0)$, where q is a positive integer, then the union of q polygons congruent to F has the same defect as F_0. By Theorem 12, this union of polygons is a polygon G equidecomposable with F_0, and hence the function s^* (satisfying axioms (β) and (γ^*)) takes the *same* value on both G and F_0. It follows that $s^*\,(F) = \frac{1}{q}\,s^*\,(F_0)$. Thus if $D\,(F) = \frac{1}{q}\,D\,(F_0)$, then $s^*\,(F) = \frac{1}{q}\,s^*\,(F_0)$. From this it follows that if $D\,(F) = \frac{p}{q}\,D\,(F_0)$, then $s^*\,(F) = \frac{p}{q}\,s^*\,(F_0)$.

We now choose an $\varepsilon > 0$ such that $(\pi - \varepsilon)\,D\,(F_0)/s^*\,(F_0) > \pi$ (ε exists, since $D\,(F_0)/s^*\,(F_0) > 1$). We then choose a triangle F, contained in the infinite "triangle," such that $s^*(F) > \pi - \varepsilon$ (Fig. 39). Here we can assume (by enlarging the triangle F, if necessary) that $D\,(F) = \frac{p}{q}\,D\,(F_0)$, where p and q are positive integers. Then, as proved above, $s^*\,(F) = \frac{p}{q}\,s^*\,(F_0)$, and therefore

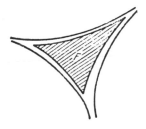

FIG. 39

$$D\,(F) = \frac{p}{q}\,D\,(F_0) = \frac{s^*\,(F)}{s^*\,(F_0)}\,D\,(F_0) > (\pi - \varepsilon)\,\frac{D\,(F_0)}{s^*\,(F_0)} > \pi,$$

which is impossible. The uniqueness now follows by contradiction.

The corollary just proved shows that in Theorem 12 we are actually talking about polygons with the same *area*, i.e., *any two polygons with the same area are equidecomposable, in both hyperbolic and elliptic geometry.* Thus equality of area and equidecomposability are *equivalent* concepts in these geometries. Equicomplementability is also equivalent to equality of area and equidecomposability (Theorem 11). Thus, unlike the Archimedean axiom, the postulate of parallels is not essential for the validity of the Bolyai–Gerwien theorem.

§9. Equidecomposability with respect to the group of translations and central inversions

The material in §§7, 8 represents, in a general way, what Hilbert meant in saying "Now the analogous problem in the plane has been solved." But Hilbert was not entirely correct. In the context of the ordinary (Euclidean) meaning of equality of area, equidecomposability and equicomplementability, his sentence is completely justified (as we saw in § 7). However, new directions of research arise in studying the connection between these concepts in *different* geometries (in particular, in non-Archimedean geometries, whose study was initiated by Hilbert himself). One of these new directions was opened by the work of the Swiss school of geometers, headed by H. Hadwiger,[9] and the results obtained in this direction will be discussed later.

An interesting problem arises in connections with the Bolyai-Gerwien theorem, namely that of imposing extra restrictions on the number or on the arrangement of the pieces making up the polygons of equal area. To explain one such restriction, imagine a plane in the form of a sheet of colored paper, with one side red and the other side (the "wrong" side) white. If two polygons of equal area are cut out of this sheet of paper, then the question arises of whether or not one of them can be cut into pieces which can be reassembled to from a *red* polygon congruent to the other one (i.e., the pieces are allowed to be moved around, provided that they are not turned upside down, letting the wrong, white side appear). The corresponding mathematical problem consists in decomposing the polygons F and G into pieces which can be obtained from each other by making *orientation-preserving* motions (i.e., rotations or translations), so that

$$F = M_1 + \ldots + M_k, \quad G = f_1(M_1) + \ldots + f_k(M_k),$$

where $f_1, \ldots, f_k$ are orientation-preserving motions.

A special case of this problem was proposed at one of the Moscow Mathematical Olympiads, in the following facetious form: An eccentric pastry chef baked a cake in the form of a *scalene* triangle. A box was also made for the cake, but by oversight it was pasted together incorrectly, so that the cake and the box were *symmetric* to each other. It is necessary to cut the cake (as economically as possible) into pieces which can be put into the box, where, naturally, pieces of cake must not be put into the box with the icing side

---

[9] Incidentally, Hadwiger has written the author a letter in which he talks about the "final" solution of the planimetric problem of equidecomposability, contained in reference 5 (the discussion was about Theorem 17, proved below). Evidently everything depends on one's point of view. In §7 we presented a solution within the framework of Euclidean conceptions, and in this and the next two sections the same problem is considered from the standpoint of Kleinian ideas.

down. Figure 40 shows a way of cutting the cake into pieces, each of which has an axis of symmetry, in a way allowing them to be put into the box.

It is now clear that the question posed above has an affirmative answer: *The equidecomposability of two polygons of equal area can be established by using orientation-preserving motions.* In fact, by the Bolyai–Gerwien theorem, we can decompose two polygons of equal area into mutually congruent pieces and hence into mutually congruent triangles. But two congruent triangles can either be obtained from each other by an orientation-preserving motion, or else (as in Fig. 40), each of them can be cut up into three pieces which can be reassembled to give the other, by making orientation-preserving motions.

The following interesting result, involving the imposition of extra requirements on the way the pieces are arranged, was obtained in 1951 by Hadwiger and Glur [30] : *The equidecomposability of two polygons of equal area can be proved by using decompositions such that the corresponding pieces have parallel sides.* At first glance, this result seems even implausible. Indeed, it is hard to believe that two congruent triangles which have been rotated relative to each other through an arbitrary angle (Fig. 41) can always be decomposed into congruent pieces whose corresponding sides are parallel.

Before proving the Hadwiger–Glur theorem, we consider the connection between the extra conditions being imposed and groups

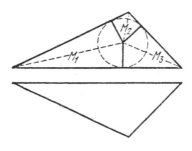

FIG. 40

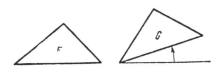

FIG. 41

of motions. Let D denote the set of all motions of the plane (both those preserving orientation and those changing it). It will be recalled that a nonempty subset G of the set D is called a *group of motions* if it has the following two properties:

1) If f and g belong to G, then so does the composition $g \circ f$ (i.e., the result of consecutively applying f and g);
2) If a motion f belongs to G, then so does the inverse motion f^{-1}.

The set D itself is obviously a group of motions. The set D_0 consisting of all *orientation-preserving* motions is also a group of motions. In fact, a composition of orientation-preserving motions also preserves orientation, and if f preserves orientation, so does f^{-1}.

The set T of all translations of the plane is a group of motions. In fact, a composition of translations is itself a translation, and the motion inverse to a translation is also a translation.

A more complicated example of a group of motions is the group S consisting of all translations and all central inversions. To verify that S is a group of motions, we take an arbitrary vector $a \neq 0$. This vector goes into itself under a translation and into $-a$ under a central inversion. Therefore, under the composition of two motions f, $g \in S$, the vector a goes either into itself or into the vector $-a$. But an orientation-preserving motion carrying a into itself is a translation, while an orientation-preserving motion carrying a into $-a$ is a central inversion. This means that a composition of motions belonging to the set S also belongs to S. Moreover, if f belongs to S, then so does f^{-1}. Thus S is a group of motions. We observe that

$$T \subset S \subset D_0 \subset D.$$

We now recall the basic tenets of Klein's geometrical outlook [41]. Let G be a group of motions of the plane.[10] A figure A is said to be G-congruent to a figure B if there exists a motion $f \in G$ such that $f(A) = B$, i.e., if A can be "superimposed" on B with the help of some motion $f \in G$. It is easy to see that if a figure A is G-congruent to a figure B, then B is also G-congruent to A. In fact, if $f(A) = B$, $f \in G$, then $f^{-1}(B) = A$, where $f^{-1} \in G$ by the definition of a group of motions. Thus $A \underset{G}{\cong} B$ implies $B \underset{G}{\cong} A$ (where the symbol $\underset{G}{\cong}$ denotes G-congruence), i.e., the relation of G-congruence is *symmetric*. It is also easy to see that the relation of G-congruence is *transitive*. In fact, let $A \underset{G}{\cong} B$ and $B \underset{G}{\cong} C$, so that there exist motions $f, g \in G$ such that $f(A) = B$, $g(B) = C$. Then

$$(g \circ f)(A) = g(f(A)) = g(B) = C,$$

i.e., the motion $g \circ f$ (which also belongs to the group G) carries the figure A into C, and hence $A \underset{G}{\cong} C$. Finally, the relation of G-congruence is *reflexive*. In fact, if f is an arbitrary motion belonging to the group G, then $f^{-1} \in G$, and hence G contains the *identity* motion. Since $e(A) = A$ for every figure A, we have $A \underset{G}{\cong} A$. Thus the relation of G-congruence is reflexive, symmetric and transitive, i.e., G-congruence is an equivalence relation.

Moreover, a property of figures is said to be *invariant under the group* G (more concisely, *G-invariant*) if it is preserved under all motions belonging to the group G (i.e., if the fact that A has the property implies that every figure G-congruent to A also has the property). For example, the property of "being a polygon" is

---

[10] We might have considered any *group of transformations* G of an arbitrary set M, but for our purposes we need only consider the case where M is the Euclidean plane, and all the transformations in the group G are motions.

G-invariant; it is preserved under *arbitrary* motions, and in particular under motions belonging to the group G. We can also talk about a property of a *pair* of figures. Thus "equality of area" is a property of a pair of figures, i.e., two figures A and B have the property if $s(A) = s(B)$. This property is also G-invariant. In fact, equality of area is preserved under *arbitrary* motions, and hence under the motions belonging to the group G as well. Of course, a group of motions G can have its own specific G-invariant properties, which are preserved under the motions belonging to G, but *not* under arbitrary motions.

According to Klein, every group of motions G defines *its own* geometry, which we call *G-geometry*. The object of study in G-geometry is the set of all possible G-invariant properties.

For example, according to Klein, Euclidean planimetry (which studies the properties of figures under arbitrary motions) is just D-geometry. Klein's scheme also comprises non-Archimedean geometry (here we must consider the group G of all motions of the non-Archimedean plane M, and study all G-invariant properties of figures lying in M), as well as hyperbolic and elliptic geometries, affine and projective geometries, and many others. As still another example, we note that T-geometry is essentially the same as vector algebra, for the fact that two vectors (directed line segments) $\overrightarrow{ab}$ and $\overrightarrow{cd}$ are equal means that they can be made to coincide (with direction taken into account) by making some translation, i.e., the directed line segments are T congruent. From now on, we will only consider groups of motions of the Euclidean plane (leaving aside the non-Archimedean plane, the hyperbolic plane, etc.).

Each G-geometry has *its own* concept of equidecomposability: Two polygons A and B are said to be *G-equidecomposable* $(A \underset{G}{\sim} B)$ if there exist polygons $A_1, \ldots, A_k$ and $B_1, \ldots, B_k$ such that

$$A = A_1 + \ldots + A_k, \quad B = B_1 + \ldots + B_k;$$

$$A_1 \underset{G}{\cong} B_1, \ldots, \quad A_k \underset{G}{\cong} B_k.$$

G-equicomplementability is defined analogously. Thus in every

G-geometry the problem arises of the equivalence of the concepts of equality of area, G-equidecomposability and G-equicomplementability.

The extra restrictions considered at the beginning of this seciton on the arrangement of the pieces making up polygons of equal area are intimately related to the concept of G-equidecomposability. The first restriction (in which only orientation-preserving motions are considered) obviously means that we are talking about D_0-equidecomposability. Thus, in connection with Fig. 40, we have already proved

Theorem 13. *Any two polygons with the same area are D_0-equidecomposable.*

Moreover, if two polygons are obtained from each other with the help of a translation or a central inversion, then their corresponding sides are parallel. Conversely, if the corresponding sides of two congruent polygons are parallel, then the polygons can be obtained from each other with the help of a translation or a central inversion. In other words, the corresponding sides of two congruent polygons are parallel if and only if the polygons are *S-congruent.* Hence the above-mentioned Hadwiger–Glur theorem can be stated in the following equivalent form:

Theorem 14. *Any two polygons with the same area are S-equidecomposable.*

This is the form in which we will prove the theorem. As in the proof of the Bolyai–Gerwien theorem, we first need a few lemmas.

Lemma 1$_G$. *If $A \underset{G}{\sim} B$ and $B \underset{G}{\sim} C$, then $A \underset{G}{\sim} C$.*

The proof is virtually the same as in the case of Lemma 1. We need only assume that the motions $f_1, \ldots, f_h, g_1, \ldots, g_l$ belong to the group G, so that the motions $g_j \circ f_i^{-1}$ also belong to G.

In particular, Lemma 1 is valid for the group S. Thus if $A \underset{S}{\sim} B$ and $B \underset{S}{\sim} C$, then $A \underset{S}{\sim} C$.

Lemma 2$_S$. *Every triangle is S-equidecomposable with some rectangle.*

In fact, this is exactly what was established in the proof of Lemma 2, since the triangles marked with the same numbers in Fig. 28 can be

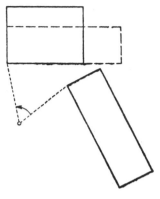

FIG. 42

obtained from each other with the help of a central inversion (reflection in a point).

Lemma 3$_T$. *Any two rectangles with the same area are T-equidecomposable.*

Here the argument used to prove Lemma 3 must be refined. In fact, the phrase "we arrange" (in the first sentence of the proof of Lemma 3) actually alludes to a rotation (Fig. 42). Since only translations of the pieces are used in the rest of the argument, it suffices to prove that if M is an arbitrary rectangle and L is a line, then there exists a rectangle N, with one side parallel to L, which is T-equidecomposable with M. Through a vertex of M we draw a line parallel to L. This line divides M into two pieces, one of which is a triangle (Fig. 43). Cutting off this triangle and using a translation to move it into a new position, we get a parallelogram, one side of which is parallel to the line L. From the vertex of an acute angle of this parallelogram we drop an altitude onto the side parallel to L. If this altitude lies entirely inside the parallelogram, it cuts off a triangle, which can then be shifted to give the required rectangle (Fig. 44). However, if the altitude lies partly outside the parallelogram, we first draw lines parallel to L decomposing the parallelogram into several congruent pieces, and then shift these pieces (Fig. 45) to

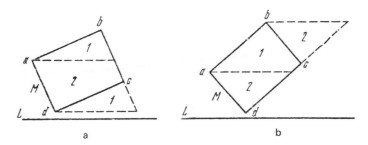

FIG. 43

obtain a parallelogram which is T-equidecomposable with the original one and which now completely contains its altitude. This completes the proof of Lemma 3_T.

We observe that if $H \supset G$, then G-equidecomposable polygons are also H-equidecomposable. In particular, the T-equidecomposability of two polygons implies their S-equidecomposability. Thus, by the lemma just proved, *any two rectangles with the same area are S-equidecomposable.*

Theorem 14 is now proved in virtually the same way as the Bolyai–Gerwien theorem. In fact, it is only necessary to change everywhere the word "equidecomposable" to "S-equidecomposable."

Finally, we note that Theorem 11 and its proof remain valid in G-geometry, i.e., *if two figures are G-equidecomposable, then they are G-equicomplementable.* Therefore, it follows from Theorem 14 that the concepts of equality of area, S-equidecomposability and S-equicomplementability are equivalent (cf. the end of §7).

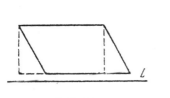

FIG. 44

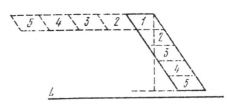

FIG. 45

§10. Equidecomposability with respect to the group of translations

Theorem 14 is the strongest of the Theorems 10, 13, and 14, since $D \supset D_0 \supset S$. Thus the following question arises quite naturally: Can a group of motions G be found, which is *smaller* than S, but still has the property that any two polygons of equal area are G-equidecomposable? In particular, is T such a group, i.e., can any two polygons of equal area be decomposed into pieces which can be obtained from one another by making translations only?

This question has a negative answer, as also shown by Hadwiger and Glur [30]. In fact, these authors found a necessary and sufficient condition for T-equidecomposability of polygons. The present section is devoted to an exposition of these results.

Given a group of motions G, a function $\varphi(F)$, defined on the set of all polygons, is said to be an *additive G-invariant* if 1) it is additive, i.e., satisfies axiom (β), so that $\varphi(F) = \varphi(F') + \varphi(F'')$ whenever $F = F' + F''$, and 2) its values are invariant under motions of the group G, i.e., $\varphi(F) = \varphi(f(F))$ for an arbitrary polygon F and an arbitrary motion $f \in G$.

Theorem 15. *Let the function $\varphi(F)$ be an additive G-invariant. Then a necessary condition for G-equidecomposability (and also for G-equicomplementability) of two polygons A and B is that $\varphi(A) = \varphi(B)$.*

PROOF. Let $A \underset{G}{\sim} B$, i.e.,

$$A = A_1 + \ldots + A_k, \quad B = B_1 + \ldots + B_k;$$
$$A_1 \underset{G}{\cong} B_1, \ldots, A_k \underset{G}{\cong} B_k.$$

By the additivity of the function $\varphi(F)$, we have

$$\varphi(A) = \varphi(A_1) + \ldots + \varphi(A_k),$$
$$\varphi(B) = \varphi(B_1) + \ldots + \varphi(B_k),$$

while

$$\varphi(A_1) = \varphi(B_1), \ldots, \varphi(A_k) = \varphi(B_k)$$

by the G-invariance of φ (F). The validity of the formula φ $(A) = \varphi$ (B) follows at once, and the proof for the case of equicomplementability is similar.

We now construct additive T-invariants, which allow us to obtain a necessary and sufficient condition for T-equidecomposability. Given a line p, we say that p is "*rigged*" if we know which of the two half-planes determined by p is regarded as *positive*. Let F be an arbitrary polygon, and let ab be one of its sides. The side ab is assigned the coefficient $\varepsilon = 0$ if it is not parallel to the line p. On the other hand, if $ab \| p$ and the polygon F adjoins ab from the *positive* side (Fig. 46), we assign ab the coefficient $\varepsilon = 1$, while if $ab \| p$ and F adjoins ab from the *negative* side, we assign ab the coefficient $\varepsilon = -1$. By the *weight* of the side ab in the polygon F we mean the number εl where l is the length of the side ab. Finally, let J_p (F) denote the sum of the weights of all the sides of the polygon F. Note that only the lengths of the sides parallel to the line p enter the sum J_p (F) (with coefficients ± 1).

Next we show that the function J_p (F) is an additive T-invariant. To this end, let $F = F' + F''$ and consider all the points which are vertices of the three polygons. These points divide the sides of the polygons F, F', and F'' into segments (in general, shorter than the sides), which we call *links*. For example, the side ab of the polygon F in Fig. 47 consists of three links am, mn, and nb. It is clear that in calculating the values of J_p (F), J_p (F') and J_p (F''), we can take the sum (with appropriate coefficients) of the lengths of the *links*,

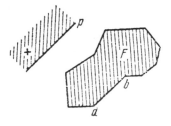

FIG. 46

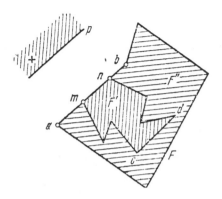

FIG. 47

rather than of the sides, since the length of each side equals the sum of the lengths of the links of which it consists.

Consider a link, like the link cd in Fig. 47, which lies entirely inside F (with the possible exception of its end points). *Both* polygons F' and F'' adjoin this link, but from different sides. Therefore, this link appears with one coefficient in calculating the sum $J_p (F')$, and with the opposite coefficient in calculating $J_p (F'')$, so that these terms contribute zero to the total sum $J_p (F') + J_p (F'')$. Thus, in calculating the sum $J_p (F') + J_p (F'')$, there is no need to consider links lying *inside F*.

Now consider a link, like the link am in Fig. 47, which lies on the boundary of the polygon F and is parallel to the line p. Only one of the polygons F', F'' adjoins this link, and it does so from the *same* side as the polygon F. Therefore, this link enters the sum $J_p (F') + J_p (F'')$ with the same sign as it enters the sum $J_p (F)$. These considerations prove the formula

$$J_p (F) = J_p (F') + J_p (F''),$$

thereby establishing the *additivity* of the function J_p. The T-invariance of J_p is obvious. Thus J_p is an additive T-invariant.

With the help of this invariant, it is easy to show that there exist polygons of equal area which are not T-equidecomposable. In fact, let A be a triangle and p a rigged line parallel to one of the sides, and let B be a parallelogram with the same area as A. It is clear that $J_p(A) \neq 0$, while $J_p(B) = 0$, so that

$$J_p(A) \neq J_p(B).$$

Hence, by Theorem 15, the polygons A and B are not T-equidecomposable.[11]

A natural problem now arises, namely that of finding *conditions* which (together with equality of area) must be imposed on the polygons A and B in order for them to be T-equidecomposable. Such conditions are contained in the following theorem due to Hadwiger and Glur [30].

Theorem 16. *A necessary and sufficient condition for two polygons A and B of equal area to be T-equidecomposable is that $J_p(A) = J_p(B)$ for every rigged line p.*

PROOF. The necessity follows from Theorem 15 (since $J_p(F)$ is an additive T-invariant). We now prove the sufficiency. Suppose two polygons of equal area have the property that $J_p(A) = J_p(B)$ for every rigged line p. We fix a line q and draw all the lines perpendicular to q going through the vertices of the polygons A and B. As a result, A and B are decomposed into pieces, each of which is either a trapezoid whose bases are perpendicular to q (with a parallelogram as a special case of a trapezoid) or a triangle one of

---

[11] We observe that with the help of the function $J_p(F)$, it is easy to establish that axiom (α) is independent of axioms (β), (γ), (δ) (for polygosn). In fact, since $J_p(F) = 0$ for an arbitrary parallelogram F, and, in particular, for the unit square Q, the function

$$s_\alpha(F) = s(F) + J_p(F),$$

considered on the set of all polygons, satisfies axioms (β), (γ), (δ) (for an arbitrary choice of the rigged line p). But s_α does not satisfy axiom (α). For example, if F is a triangle of area $1/2$, with a side of length 1 parallel to p, which it adjoins from the *negative* side, then

$$s_\alpha(F) = s(F) + J_p(F) = 1/2 - 1 < 0.$$

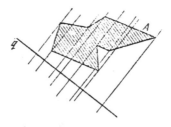

FIG. 48

whose sides is perpendicular to q (Fig. 48). The triangle can be replaced by a sum of two trapezoids (Fig. 49), while the trapezoid can be decomposed into *rectangular* trapezoids (to achieve this, it may be necessary to dissect the trapezoid into "narrower" trapezoids by drawing lines parallel to the bases, as in Fig. 50). Thus

$$A \underset{T}{\sim} A_1 + \ldots + A_h, \quad B \underset{T}{\sim} B_1 + \ldots + B_l, \quad (11)$$

where the A_i, B_j are rectangular trapezoids whose bases are perpendicular to q. In other words, each of the trapezoids A_i, B_j has no

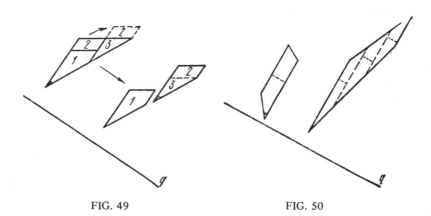

FIG. 49   FIG. 50

more than *one* side which is neither parallel nor perpendicular to q.

Suppose the trapezoid A_1 is not a rectangle, i.e., has a side a_1b_1 which is neither parallel nor perpendicular to q. Let p_1 be a rigged line parallel to a_1b_1, where we choose the positive half-plane in such a way that $J_{p_1}(A_1) > 0$, so that A_1 adjoins a_1b_1 from the positive side. Suppose $J_{p_1}(A_2) < 0$, which means that A_2 has a side a_2b_2 parallel to p_1 and adjoins a_2b_2 from the negative side. To be explicit, suppose the side a_1b_1 is no shorter than a_2b_2. Then along a_1b_1 we lay off a segment a_1c congruent to a_2b_2, and through the point c we draw a line perpendicular to q. This line divides A_1 into two trapezoids A_1' and M_1, the first of which has a side a_1c congruent to a_2b_2 (Fig. 51). By making a transition, we attach A_1' and A_2 to each other along their parallel sides, thereby obtaining a *rectangle* $M_2 \underset{T}{\sim} A_2 + A_1'$, where $A_1 + A_2 \underset{T}{\sim} M_1 + M_2$.

As a result, from *two* trapezoids A_1 and A_2 for which the invariant J_{p_1} takes values of opposite signs, we obtain the rectangle M_2 and a *single* trapezoid M_1 for which J_{p_1} takes a nonzero value. Repetition of this process leads to a new formula $A \underset{T}{\sim} M_1 + \ldots$ $+ M_k$, in which all the numbers $J_{p_1}(M_1), \ldots, J_{p_1}(M_k)$ that are nonzero have the *same* sign, and where, as before, the polygons M_1, $\ldots$, M_k are trapezoids whose bases are perpendicular to the line q

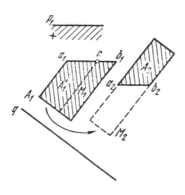

FIG. 51

(with rectangles as special cases). Similarly, we have $B \underset{T}{\sim} N_1 + \ldots$
$+ N_l$. Since $J_{p_1}(A) = J_{p_1}(B)$, it follows that

$$J_{p_1}(M_1) + \ldots + J_{p_1}(M_k) = J_{p_1}(N_1) + \ldots + J_{p_1}(N_l).$$

Therefore, all the numbers $J_{p_1}(M_i)$, $J_{p_1}(N_j)$ that are nonzero
have the same sign. To be explicit, suppose they are all positive.

Thus we can assume that $J_{p_1}(M_1) > 0, J_{p_1}(N_1) > 0$. Let $c_1 d_1$
and $c_2 d_2$ denote the sides of the trapezoids M_1 and N_1 which are
parallel to the line p_1, and suppose the side $c_2 d_2$ is no shorter than
$c_1 d_1$, say. Then along $c_2 d_2$ we lay off a segment $c_2 e$ congruent to
$c_1 d_1$, and through the point e we draw a line perpendicular to q. This
line divides N_1 into two trapezoids N_1' and N_1^*, the first of which
has a side $c_2 e$ congruent to $c_1 d_1$ (Fig. 52). To be explicit, suppose
the bases of the trapezoid M_1 are larger than the corresponding bases
of N_1'. Then (Fig. 53a)

$$M_1 = H_1 + \Pi, \quad N_1' = t_1(H_1),$$

where Π is a rectangle and t_1 a translation. (If M_1 has smaller bases,
then $M_1 = H_1$, $N_1' = t_1(H_1) + \Pi$, as in Fig. 53b, and this does

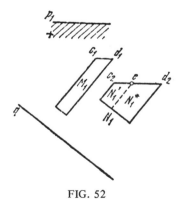

FIG. 52

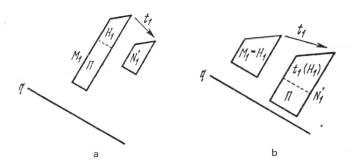

FIG. 53

not change the character of the subsequent argument). Thus we get

$$A \underset{T}{\sim} (H_1 + \Pi) + M_2 + \ldots + M_h, \quad B \underset{T}{\sim} (t_1 (H_1) + N_1^{\bullet}) +$$
$$+ N_2 + \ldots + N_l. \quad (12)$$

where the trapezoids H_1 and $t_1 (H_1)$ are T-congruent, and Π is a rectangle. Moreover, the number of remaining trapezoids $M_2, \ldots,$ $M_h, N_1^{\bullet}, N_2, \ldots, N_l$, namely $k + l - 1$, is now less than the number of trapezoids figuring in the right-hand sides of the relations (11).

Next, if $N_1^{\bullet}$ is not a rectangle, we can repeat the same construction, obtaining relations of the form

$$A \underset{T}{\sim} H_1 + H_2 + \Pi + M_2^{\bullet} + \ldots + M_h,$$
$$B \underset{T}{\sim} t_1 (H_1) + t_2 (H_2) + \Pi' + N_2 + \ldots + N_l;$$

and so on. Thus we finally arrive at the relations

$$A \underset{T}{\sim} H_1 + \ldots H_r + \Pi_1 + \ldots + \Pi_\alpha,$$
$$B \underset{T}{\sim} t_1 (H_1) + \ldots + t_r (H_r) + \Pi_1' + \ldots + \Pi_\beta',$$

where $t_1, \ldots, t_r$ are translations and the Π_i, Π'_j are rectangles.

It is now an easy matter to complete the proof. The polygons A and B have equal area, and the same is true of the trapezoids H_i and $t_i(H_i)$. Hence the polygons $\Pi_1 + \ldots + \Pi_\alpha$ and $\Pi'_1 + \ldots + \Pi'_\beta$ have equal areas. It is then an easy consequence of Lemma 3_T (p. 76) that

$$\Pi_1 + \ldots + \Pi_\alpha \underset{T}{\sim} \Pi'_1 + \ldots + \Pi'_\beta.$$

But this implies the required relation $A \underset{T}{\sim} B$.

It should be noted that although in Theorem 16 (whose proof is now complete) we talk about the formula $J_p(A) = J_p(B)$ holding for an *arbitrary* rigged line, actually we need only verify that $J_p(A) = J_p(B)$ for a *finite* number of lines, namely the lines parallel to the sides of the polygons A and B (since $J_p(A) = J_p(B) = 0$ for all the other rigged lines). Thus the condition given in the statement of Theorem 16 is completely practical.

As an application of Theorem 16, consider the problem of finding the *convex* polygons which are T-equidecomposable with a square. For a square Q we have $J_p(Q) = 0$ (for every rigged line p). Therefore, by Theorem 16, the problem reduces to finding the convex polygons for which the invariant J_p equals zero for an arbitrary rigged line p. Suppose a polygon F has this property, and let p be a rigged line parallel to the side ab of F. Then the polygon F must have another side parallel to ab, say mn, where the sides ab and mn must be congruent (since otherwise the number $J_p(F)$, equal to the difference between the lengths of the sides ab and mn, would not equal zero). Thus for every side of the polygon F, there is a congruent side parallel to the given side. But then, as is easily verified, F is centrally symmetric.[12] It is clear that the converse is also true, namely, if F is centrally symmetric, then $J_p(F) = 0$ for every rigged line. Thus *a convex polygon F is T-equidecomposable with the square of the same area if and only if the polygon is centrally symmetric* [30].

---

[12] That is, symmetric with respect to central inversion (reflection in a point).

§11. Minimality of the group of translations and central inversions

We now answer the question posed at the beginning of the preceding section, i.e., we prove that S is the *smallest* group of motions leading to a proof of the equidecomposability of arbitrary polygons of equal area.

Theorem 17. (see [5], [6]). *If G is a group of motions such that any two polygons of euqal area are G-equidecomposable, then $G \supset S$; i.e., G contains all translations and central inversions.*

The proof depends on two lemmas. In the statement of these lemmas, we will always assume that G is a group satisfying the conditions of the theorem.

Lemma 7. *The group G is transitive, i.e., given any two points p and q of the plane, there is a motion in the group G carrying p into q.*

Suppose, to the contrary, that there exist points p and q such that no motion belonging to the group G carries the point p into the point q, and let W denote the set of all points into which the point p can be carried by a motion belonging to G, so that $q \notin W$. Given an arbitrary polygon F, let $I_W(F)$ denote the sum of the angles at the vertices of F which belong to the set W. Then, as we now show $F = F_1 + \ldots + F_k$, implies

$$I_W(F) = I_W(F_1) + \ldots + I_W(F_k) + n\pi, \qquad (13)$$

where n is an integer. It suffices to prove this for $k = 2$, i.e., for $F = F_1 + F_2$. Every vertex a of the polygon F is either a common vertex of the polygons F_1 and F_2 (Figs. 54a, b), or a vertex of one of them (say F_1), where in this case a can be an *interior* point of a side of the polygon F_2 (Fig. 54c) or may not belong to F_2 at all (Fig. 54d). In the first case (Fig. 54a, b) we have $\alpha = \alpha_1 + \alpha_2$, where $\alpha, \alpha_1, \alpha_2$ are the vertex angles of the polygons F, F_1, F_2 at the vertex a, while in the second case $\alpha = \alpha_1 + \pi$ (Fig. 54c) or $\alpha = \alpha_1$ (Fig. 54d). There is a similar equation at every vertex of F belonging to the set W, and adding them all up, we get formula (13).

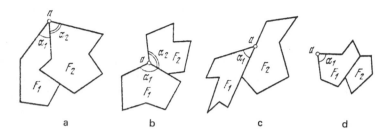

FIG. 54

Moreover, the function I_W is G-invariant, i.e.,

$$I_W(F) = I_W(g(F)) \tag{14}$$

for every polygon F and every motion $g \in G$. In fact, if a is a vertex of the polygon F, then the corresponding vertex $g(a)$ of the polygon $g(F)$ belongs to W if and only if $a \in W$ and the angles of the polygons F and $g(F)$ at the corresponding vertices a and $g(a)$ are equal, which implies (14). It follows from (13) and (14) that

$$I_W(M) = I_W(N) + l\pi \quad \text{if} \quad M \underset{G}{\sim} N, \tag{15}$$

where l is some integer.

Now let pqm and pqn be two congruent obtuse isosceles triangles, where the first triangle has the vertex of its obtuse angle at p, and the second triangle has it at q. Let the acute angle of these triangles be denoted by α. Since $p \in W$, $q \notin W$, the function I_W takes either the value $\pi - \alpha$ on the triangle pqm or the value $\pi - 2\alpha$ (depending on whether or not m belongs to W). On the other hand, the value I_W on the triangle pqn equals α or 2α. Therefore, since $\alpha < \pi/4$, the equality $I_W(\triangle\, pqm) = I_W(\triangle\, pqn) + l\pi$ cannot hold for any integer l, and hence, by (15), the tirangles pqm and pqn cannot be G-equidecomposable. But this contradicts the properties of the group G, thereby proving the transitivity of G, since polygons of equal area (and hence, *a fortiori*, congruent polygons) are necessarily G-equidecomposable.

Remark. Formulas (13) and (15) can be written somewhat differently. Let X denote the set of all numbers of the form $l\pi$, where l is an integer. The set X is a *subgroup* of the additive group R of all real numbers. Let φ denote the natural homomorphism of the group R onto the factor group $R_\pi = R/X$. Then $\varphi(l\pi) = 0$ for every integer l, and hence formulas (13)–(15) become

$$\varphi(I_W(F)) = \varphi(I_W(F_1)) + \ldots + \varphi(I_W(F_k)), \qquad (13')$$

$$\varphi(I_W(F)) = \varphi(I_W(g(F))), \qquad (14')$$

$$\varphi(I_W(M)) = \varphi(I_W(N)) \quad \text{if} \quad M \underset{G}{\sim} N. \qquad (15')$$

Formulas $(13')$ and $(14')$ show that $\varphi(I_W(F))$ is an additive G-invariant with values in R_π, and with this treatment, formula $(15')$ is a consequence of Theorem 15.

Lemma 8. *The group G contains at least one central inversion.*

Every motion of the plane is either (see [56]) a translation, a rotation, or a glide reflection (i.e., the composition of a symmetry with respect to some line and a translation along the line, called the *axis* of the glide reflection). Suppose G contains at least one glide reflection h. Then we fix any directed line p perpendicular to the axis of h, and observe that the motion h carries p into the line with the *opposite* direction. However, if the group G does not contain any glide reflections, we fix an *arbitrary* directed line p. Moreover, we agree to regard a line p' as "*marked*" if there is an orientation-preserving motion in the group G carrying p into a line which is parallel to p' and has the same direction as p'.

Suppose that given any two oppositely directed lines, no more than *one* of them is marked. Consider an arbitrary polygon F, and let ab be a side of F. On the line ab we choose a direction such that the polygon F adjoins ab from the *left*. If the line ab with this direction is marked, then the side ab is assigned the coefficient $\varepsilon = 1$, while if the oppositely directed line is marked, we assign ab the coefficient $\varepsilon = -1$. On the other hand, if neither of these lines is marked, we assign ab the coefficient $\varepsilon = 0$. Let $K_p(F)$ denote the sum of the lengths of the sides of the polygon F, each multiplied by the appropriate value of the coefficient ε.

The function K_p (F) constructed in this way is an additive G-invariant. The proof of the additivity is virtually the same as in the case of the function J_p (F) (p. 79). We now prove the G-invariance. To this end, let F be an arbitrary polygon, and let $g \in G$. Moreover, let $a_1 b_1$ be a side of the polygon F, while $a_2 b_2$ is the corresponding side of the polygon g (F), i.e., $a_2 = g$ (a_1), $b_2 = g$ (b_1). Let p_1 denote the line $a_1 b_1$, directed in such a way that the polygon F adjoins the side $a_1 b_1$ from the left, let p_2 be the directed line into which p_1 is carried by the motion g, and let p_2' be the line p_2 taken with the opposite direction. Suppose the side $a_1 b_1$ of the polygon F is assigned the coefficient $\varepsilon = 1$, i.e., suppose p_1 is marked. Then there exists an orientation-preserving motion $f \in G$ carrying p into a line which is parallel to p_1 and has the same direction as p_1. If the motion g is also *orientation-preserving,* then $g \circ f$ is an orientation-preserving motion carrying p into a line which is parallel to p_2 and has the same direction as p_2, i.e., the line p_2 is also marked. Since g is orientation-preserving, it follows that g (F) adjoins the segment $a_2 b_2$ of the directed line p_2 from the *left* (in the same way as F adjoins $a_1 b_1$), i.e., the side $a_2 b_2$ of the polygon g (F) is assigned the coefficient $\varepsilon = 1$. However, if the motion g *changes* the orientation, then g (F) adjoins the side $a_2 b_2$ from the left only if we assign $a_2 b_2$ the direction p_2' (rather than p_2). Then the motion $g \circ f \circ h$ is orientation-preserving and carries p into a line which is parallel to p_2 and has the same direction as p_2'. Hence the line p_2' is marked, i.e., the side $a_2 b_2$ of the polygon g (F) is assigned the coefficient $\varepsilon = 1$ in this case as well.

Thus, if the side $a_1 b_1$ of the polygon F is assigned the coefficient $\varepsilon = 1$, the same is true of the side $a_2 b_2$ of the polygon g (F). In the same way, it can be shown that if the side $a_1 b_1$ is assigned the coefficient $\varepsilon = -1$, the same is true of the side $a_2 b_2$. Finally, if the side $a_1 b_1$ is assigned the coefficient $\varepsilon = 0$, the same is true of the side $a_2 b_2$ (in fact, the polygon F is obtained from g (F) by making the motion g^{-1}, and if the side $a_2 b_2$ were assigned the coefficient $\varepsilon = \pm$ 1, then the same would be true of the side $a_1 b_1$). Thus corresponding sides of the polygons F and g (F) appear in the sums K_p (F) and K_p $(g$ $(F))$ with the same coefficients, so that K_p $(F) = K_p$ $(g$ $(F))$. and the G-invariance of K_p (F) is proved.

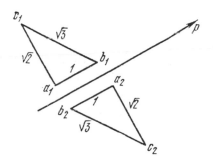

FIG. 55

Now consider two triangles $a_1 b_1 c_1$ and $a_2 b_2 c_2$ with side lengths 1, $\sqrt{2}$, $\sqrt{3}$, arranged as shown in Fig. 55. Since the triangles are congruent (and hence of equal area), they are G-equidecomposable. Therefore, by Theorem 15, $K_p\,(\,a_1 b_1 c_1) = K_p\,(a_2 b_2 c_2)$, i.e.,

$$\alpha_1 \cdot 1 + \beta_1 \cdot \sqrt{2} + \gamma_1 \cdot \sqrt{3} = \alpha_2 \cdot 1 + \beta_2 \cdot \sqrt{2} + \gamma_2 \cdot \sqrt{3},$$

where α_1, β_1, γ_1 are the coefficients assigned to the sides of the triangle $a_1 b_1 c_1$, and α_2, β_2, γ_2 are those assigned to $a_2 b_2 c_2$. Since all the coefficients are integers $(+1, -1$ or $0)$, the equation just written can hold only if $\alpha_1 = \alpha_2$, $\beta_1 = \beta_2$, $\gamma_1 = \gamma_2$. But it is clear that $\alpha_1 = 1$, $\alpha_2 = -1$, i.e., $\alpha_1 \neq \alpha_2$. Thus the assumption that given any two oppositely directed lines, no more than one of them is marked, leads to a contradiction. Therefore, we can find two oppositely directed lines q_1 and q_2 which are both marked. Let $g_i \in G$ $(i = 1, 2)$ denote the orientation-preserving motion carrying the directed line p into a line which is parallel to q_i and has the same direction as q_i. Then $g_2 \circ g_1^{-1} \in G$ is an orientation-preserving motion which carries q_1 into the oppositely directed line q_2, i.e., $g_2 \circ g_1^{-1}$ is a central inversion. The proof of Lemma 8 is now complete.

Proof of Theorem 17. Let r be a central inversion belonging to the group G (Lemma 8), let o_1 be the center of this inversion, and let o be an arbitrary point of the plane. Moreover, let $g \in G$ be the motion carrying o into o_1 (Lemma 7). Then the motion $f = g^{-1} \circ (r \circ g)$ belongs to the group G and is a central inversion. But f leaves the point o fixed, as is easily verified, and hence f is the central inversion with respect to o. Thus the central inversion with respect to an arbitrary point o belongs to the group G, i.e., G contains all central inversions. Moreover, since an arbitrary translation is a composition of two central inversions, G also contains all translations. Therefore $G \supset S$.

CHAPTER 3

EQUIDECOMPOSABILITY OF POLYHEDRA

§12. Equidecomposability of symmetric polyhedra

In the passage quoted at the end of §6, Hilbert says that "Gerling also succeeded in proving the equality of symmetrical polyhedra by dividing them into congruent parts." From this it is clear that Hilbert does not regard symmetric polyhedra as congruent. In three-dimensional Euclidean space reflection in a plane (as well as central inversion, i.e., reflection in a point) is a motion which *changes* orientation, but the only polyhedra (or, in general, figures) which Hilbert regards as congruent are those that can be obtained from each other with the help of motions that *preserve* orientation.

The meaning of Gerling's theorem (proved by him in 1844, and proved again in 1896 by Bricard [9]) is best explained from a group-theoretic point of view. We will consider various *groups of motions* of three-dimensional Euclidean space R^3, namely, the group D of all motions, the group D_0 of all orientation-preserving motions, the group T of all translations, and others. The concepts of G-congruence, G-equidecomposability and G-equicomplementability (where G is a group of motions of the space R^3) are defined for polyhedra in the same way as for polygons in the plane.

Thus, by "congruence" Hilbert means only D_0-congruence, and accordingly, when he talks about proving the equality of volumes by making a decomposition into "congruent parts," he has D_0-equidecomposability in mind. In other words, the content of Gerling's theorem is that *symmetric polyhedra are D_0-equidecomposable*. It is clear that symmetry is not the essential thing here: We are really talking about polyhedra obtained from each other with the help of a motion that *changes* orientation. For if two polyhedra are obtained from each other with the help of a motion that *preserves* orientation, then they are not only D_0-equidecomposable but actually D_0-congruent. Hence Gerling's theorem can be formulated as follows:

Theorem 18. *If two polyhedra are D-congruent, then they are D_0-equidecomposable.*

PROOF. If M is a polyhedron symmetric with respect to some plane α, and N is a polyhedron D-congruent to M, then M and N are also D_0-congruent. In fact, if f is a motion carrying M into N, and s is reflection in the plane α, then the motion $f \circ s$ also carries M into N. But one of the motions f and $f \circ s$ preserves orientation, i.e., belongs to the group D_0.

It follows from this that if a polyhedron A can be decomposed into pieces $M_1, \ldots, M_h$, each of which has a plane of symmetry, and if B is any polyhedron D-congruent to A, then B is D_0-equidecomposable with A. In fact, decomposing B into pieces $N_1, \ldots, N_h$ which are D-congruent to the polyhedra $M_1, \ldots, M_h$ respectively, we find (because of the symmetry of the polyhedron M_i) that M_i and N_i are not only D-congruent, but also D_0-congruent. Therefore, A and B are D_0-equidecomposable.

Thus it suffices to show that any polyhedron A can be decomposed into pieces, each of which has a plane of symmetry. Let A be an arbitrary polyhedron, and draw all the planes containing its faces. These planes divide A into a finite number of *convex* polyhedra. Moreover, each convex polyhedron can be decomposed into (polygonal) pyramids. In fact, we need only choose a point q inside the polyhedron, and then consider all the pyramids which have q as vertex and the faces of the polyhedron as their bases (Fig. 56). Finally, we can decompose every polygonal pyramid into several

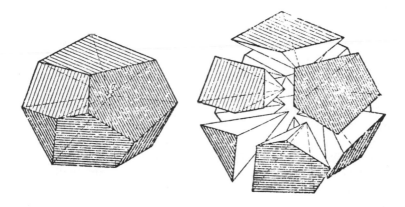

FIG. 56

triangular pyramids (Fig. 57). Hence the polyhedron A can be decomposed into triangular pyramids, and it remains to show that any triangular pyramid can be decomposed into pieces, each of which has a plane of symmetry.

Let $abcd$ be a triangular pyramid, let o be the center of the sphere inscribed in $abcd$, and let a_1, b_1, c_1, d_1 be the points at which the inscribed sphere is tangent to the faces of the pyramid (the point a_1 lies in the face that does not go through the vertex a, the point b_1 in the face that does not go through b, and so on). We now decompose the pyramid $abcd$ into the six polyhedra

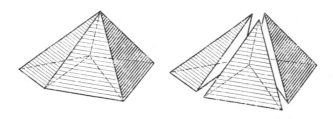

FIG. 57

$$oa_1b_1cd, \quad oa_1bc_1d, \quad oab_1c_1d, \quad oa_1bcd_1, \quad oab_1cd_1, \quad oabc_1d_1. \quad (16)$$

To complete the proof, we show that each of these polyhedra has a plane of symmetry.

The point o is at the same distance from the planes acd and bcd (since the inscribed sphere is tangent to both planes), and hence the point o lies in the plane β which goes through the edge cd and bisects the dihedral angle of the pyramid $abcd$ with cd as its edge. Therefore, the points b_1 and a_1 (i.e., the projections of the point o onto the planes acd and bcd) are symmetric with respect to the plane β, and hence the polyhedron oa_1b_1cd is also symmetric with respect to β. The symmetry of the remaining polyhedra (16) is proved similarly.

Corollary. *The concepts of D-equidecomposability and D_0-equidecomposability (for polyhedra in R^3) are equivalent.*

In fact, suppose A and B are D-equidecomposable, so that $A = M_1 + \ldots + M_k$, $B = N_1 + \ldots + N_k$, where $M_i \underset{D}{\cong} N_i$, $i = 1, \ldots, k$ (as before, $P_1 + \ldots + P_k$ means the union of polyhedra which are assumed to have pairwise disjoint interiors). By Theorem 18, the relation $M_i \underset{D}{\cong} N_i$ implies that M_i and N_i, are D_0-equidecomposable. Therefore, A and B are also D_0-equidecomposable. Conversely, if A and B are D_0-equidecomposable, they are also D-equidecomposable (since $D_0 \subset D$).

The corollary just proved shows that the solution of Hilbert's third problem does not depend on whether we interpret "equidecomposability" as D_0-equidecomposability or as D-equidecomposability, and it is the latter aspect (D-equidecomposability) that we have in mind in the next section, devoted to the solution of Hilbert's third problem. The reason why Hilbert confined himself to D_0-equidecomposability in posing the problem (by considering only orientation-preserving motions) is clear: Such a motion can be accomplished by *continuous displacement* of the space R^3 into itself as a "rigid whole," whereas a motion that changes orientation can be accomplished by continuous displacement only by going from the original space into a higher-dimensional space. This difference between motions that preserve orientation and those that change it becomes particularly apparent in the case of polygons or polyhedra in spherical spaces. For example, symmetric scalene spherical triangles (Fig. 58) cannot be made to coincide either on the sphere or in the three-dimensional space containing the sphere. In fact, making them coincide by a continuous displacement requires going from ordinary space into four-dimensional Euclidean space (or three-dimensional spherical space). For this reason, the fact that two symmetric figures on the sphere have equal area is not regarded as "obvious" (i.e., is not axiomatized), but rather is *proved* by showing that they are equidecomposable.

We note that the proof of Gerling's theorem given above generalizes verbatim to the case of polyhedra in a spherical or elliptic space (and also in a Lobachevskian space) of any number of dimensions. In fact, a polyhedron (in an n-dimensional space of constant curvature, whether Euclidean, hyperbolic or elliptic) can be decomposed into n-dimensional simplexes, and each n-dimensional simplex can be decomposed into $\frac{n(n+1)}{2}$ polyhedra which are symmetric under reflection. This decomposition is accomplished as follows, with the help of the sphere inscribed in the n-dimensional simplex: Let T be an $(n-2)$-dimensional face of the simplex, let o be the center of the sphere, and let a_1, b_1 be the points at which the

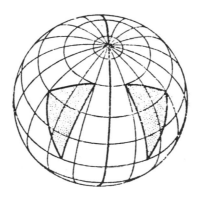

FIG. 58

sphere is tangent to the $(n - 1)$-dimensional faces of the simplex which contain T. Then the convex hull of the set $T \cup \{o, a_1, b_1\}$ is a polyhedron which is symmetric under reflection. There are a total of $\frac{n(n+1)}{2}$ such polyhedra (one for each $(n - 2)$-dimensional face), and together they make up the decomposition of the n-dimensional simplex. For $n = 3$ such a decomposition was considered in the proof of Gerling's theorem, and for $n = 2$ in the proof of Theorem 13 (see Fig. 40).

Gerling's theorem, as just presented, together with the material in §§9 and 10, gives an idea of the state of the theory of equidecomposability at the turn of the twentieth century, when Hilbert formulated his problems.[13] Individual examples of equidecomposable polyhedra were also known. The most interesting of these (apart from trivial examples, like the equidecomposability of the oblique and right prisms in Fig. 21) were found in 1896 by the English mathematician Hill [34], who gave examples of tetrahedra that are equidecomposable with a cube.

One of Hill's tetrahedra is shown in Fig. 59. Here ab, bc and cd are mutually perpendicular edges which have the same length l, while the segments bm and mn have length $l/3$. Figure 59 shows how the tetrahedron can be decomposed into four polyhedra which can be reassembled to give a right triangular prism. This prism is in turn equidecomposable with a rectangular parallelepiped, and hence also with a cube. (In §18 we will go into more detail about tetrahedra that are equidecomposable with a cube.)

Thus tetrahedra do exist whose volumes can be found by the method of decomposition. But Hilbert foresaw that Gerling's theorem, Hill's tetrahedra, and the like, are only *special*, successfully chosen examples of equidecomposable polyhedra of the same volume, which should be regarded as exceptions, rather than as illustrations of

---

[13] A theorem due to Dehn (proved below as Theorem 20) gives the solution of Hilbert's third problem. This theorem can be found in Bricard's paper [9] dated 1896 (see §15), but Bricard's proof is incorrect. Thus reference 9 actually contains no results, but it apparently influenced the work of Hilbert and Dehn.

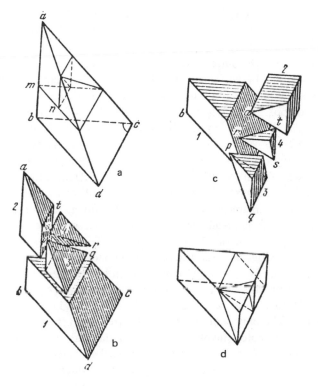

FIG. 59

a *general* law. This is what he means when he says "it seems to me probable that a general proof of this kind for the theorem of Euclid just mentioned is impossible." We have now fully explained the meaning of Hilbert's formulation of his third problem, as given at the end of §6.

§13. The solution of Hilbert's third problem

Hilbert's third problem was solved in the very year 1900 in which he read his report "Mathematical Problems." Hilbert was right: The methods of decomposition and complementation are *incapable* of proving the formula for the volume of a pyramid (in the general case). This was proved by Dehn [12], who showed that there exist polyhedra which have the same volume, but are not equidecomposable. In particular, a cube and the regular tetrahedron of the same volume are not equidecomposable (and cannot be made into equidecomposable polyhedra by the addition of congruent pieces). There also exist nonequidecomposable tetrahedra with congruent bases and the same altitude. This explains the necessity of using nonelementary methods in the theory of the volume of polyhedra.

In a way, Dehn's work has led several lives. Dehn's own exposition was hard to understand. In 1903 Kagan published a paper [39], in which Dehn's argument was considerably refined, and presented in a more systematic and readable fashion. This was, so to speak, the rebirth of Dehn's paper. In the 1950's a number of interesting results in the theory of equidecomposability were obtained by the Swiss geometer Hadwiger and his students. Their work (in particular, references [21], [24], [27]) allows one to take a new look at the work of Dehn, and to obtain Dehn's basic result by using transparent ideas in a modern treatment. The only shortcoming in this treatment is the application of the axiom of choice (involving the use of rational bases in the field of real numbers; cf. p. 29). Finally, a reworked version of Hadwiger's proof was given in reference [6], in which consideration of the whole real line is replaced by consideration of *finite* sets, which allows one to avoid the use of the axiom of choice. The proof of Dehn's theorem given in [6] is apparently the simplest, and it is this proof that will be presented here.

Given a set of real numbers M, we say that the numbers $x_1, x_2, \ldots,$ $x_k \in M$ are *linearly dependent with integral coefficients* if there exist *integers* $n_1, n_2, \ldots, n_k$, not all equal to zero, such that

$$n_1 x_1 + n_2 x_2 + \ldots + n_k x_k = 0. \tag{17}$$

The relation (17) will be called a *linear dependence with integral coefficients* or simply a *linear dependence.*[14] A real function $f(x)$,

---

[14] In §4 we considered linear dependences with rational coefficients, which do not differ in principle from linear dependences with integral coefficients (since the former can be reduced to the latter by multiplying by a common denominator of the rational coefficients).

defined on the set M, will be called *additive* if for *every* linear dependence (17) between elements of M, the same linear dependence holds between the corresponding values of f, i.e.,

$$n_1 f(x_1) + n_2 f(x_2) + \ldots + n_h f(x_h) = 0.$$

Now let A be a polyhedron, let $\alpha_1, \alpha_2, \ldots, \alpha_p$ be the dihedral angles of A expressed in radians, and let $l_1, l_2, \ldots, l_p$ be the lengths of the corresponding edges of A. If f is an additive function defined on a set M containing all the numbers $\alpha_1, \alpha_2, \ldots, \alpha_p$, then we denote the sum

$$l_1 f(\alpha_1) + l_2 f(\alpha_2) + \ldots + l_p f(\alpha_p)$$

by $f(A)$ and call it the *Dehn invariant* of the polyhedron A. (These definitions are not to be found in [12], but they stem from Dehn's work conceptually.)

The following theorem is a modification of a result due to Hadwiger [27], and represents the key to the solution of Hilbert's third problem.

Theorem 19. *Given a polyhedron A with dihedral angles $\alpha_1, \ldots, \alpha_p$ and a polyhedron B with dihedral angles $\beta_1, \ldots, \beta_q$, let M be a set of real numbers containing the numbers*

$$\pi, \alpha_1, \ldots, \alpha_p, \beta_1, \ldots, \beta_q. \tag{18}$$

Suppose there exists an additive function f defined on M such that $f(\pi) = 0$, while the corresponding Dehn invariants of the polyhedra A and B are unequal, i.e., $f(A) \neq f(B)$. Then the polyhedra A and B are nonequidecomposable (and nonequicomplementable).

The proof of Theorem 19 will be given in the next section, and we now use the theorem to present Dehn's results. In keeping with the formulation of Theorem 19, we will henceforth always assume that the additive functions under consideration satisfy the condition $f(\pi) = 0$.

Lemma 9. *The number* $\dfrac{1}{\pi}$ arc cos $\dfrac{1}{n}$ *is irrational for every integer* $n \geqslant 3$.

PROOF. Suppose to the contrary that $\varphi/\pi = l/k$, where k and l are positive integers and $\varphi =$ arc cos $1/n$. Since $k\varphi = l\pi$, we have cos $k\varphi = \pm 1$, i.e., cos k is an *integer*. We will now show that this assertion leads to a contradiction, i.e., we will show that cos $k\varphi$ cannot be an integer for any $k = 1, 2, \ldots$ Our starting point is the formula

$$\cos (k + 1) \varphi + \cos (k - 1) \varphi = 2 \cos k\varphi \cos \varphi,$$

which implies

$$\cos (k + 1) \varphi = \frac{2}{n} \cos k\varphi - \cos (k - 1) \varphi. \tag{19}$$

(bear in mind that cos $\varphi = 1/n$. The rest of the proof depends on whether n is odd or even, and we consider these two cases separately.

Suppose first that n is odd. Then, as we now show, cos $k\varphi$ can be expressed as a fraction with denominator n^k and numerator equal to an integer which is relatively prime to n; this will imply that cos $k\varphi$ cannot be an integer for any $k = 1, 2, \ldots$. For $k = 1$ and $k = 2$ this assertion can be verified directly:

$$\cos \varphi = \frac{1}{n}, \quad \cos 2\varphi = 2 \cos^2 \varphi - 1 = \frac{2}{n^2} - 1 = \frac{2 - n^2}{n^2}$$

(since n is odd, the numbers n and 2 are relatively prime, and hence so are the numbers $2 - n^2$ and n). Suppose that our assumption can be proved for all numbers $1, 2, \ldots, k$ (where $k \geqslant 2$). Then the assertion also holds for the number $k + 1$. In fact, by the induction hypothesis,

$$\cos k\varphi = \frac{a}{n^k}, \quad \cos (k - 1) \varphi = \frac{b}{n^{k-1}},$$

where a and b are integers relatively prime to n. By (19),

$$\cos(k + 1)\varphi = \frac{2}{n} \cdot \frac{a}{n^k} - \frac{b}{n^{k-1}} = \frac{2a - bn^2}{n^{k+1}}$$

and since the numbers 2 and a are relatively prime to n, the number $2a$ is relatively prime to n, and hence the same is true of the numerator $2a - bn^2$. This induction establishes the lemma for the case of odd n.

Suppose now that n is even, i.e., $n = 2m$, where $m \geqslant 2$ is an integer. Then $\cos k\varphi$ can be expressed as a fraction with denominator $2m^k$ and numerator equal to an integer which is relatively prime to m (this is proved by an analogous induction). Hence, in this case also, $\cos k\varphi$ cannot be an integer for any $k = 1, 2, \ldots$

Theorem 20 (see [12]). *A regular tetrahedron and the cube with the same volume are nonequidecomposable (and nonequicomplementable).*

PROOF. Let P be the regular tetrahedron and Q the cube with the same volume, and let φ be the dihedral angle of P. Then it is easy to see that $\varphi = \mathrm{arc\ cos}\ \frac{1}{3}$ (in Fig. 60 the point e is the centroid of the face abc, the segment de is the altitude, and the formula $|ef| = \frac{1}{3}|af| = \frac{1}{3}|df|$ holds).

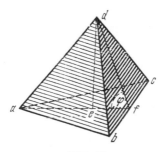

FIG. 60

We now apply Theorem 19. Each dihedral angle of the cube Q equals $\frac{\pi}{2}$, and hence in in this case the numbers (18) figuring in Theorem 19 are just π, $\frac{\pi}{2}$, φ. For M we take the set consisting of these three numbers, i.e., $M = \{\pi, \frac{\pi}{2}, \varphi\}$, and on M we define a real function f by setting

$$f(\pi) = 0, \quad f\left(\frac{\pi}{2}\right) = 0, \quad f(\varphi) = 1. \tag{20}$$

It is not hard to see that this function, defined on the set M, is *additive*. In fact, let

$$n_1\pi + n_2 \cdot \frac{\pi}{2} + n_3\varphi = 0 \tag{21}$$

be a linear dependence with integral coefficients between the elements of M. Assuming that $n_3 \neq 0$, we deduce from (21) that the number

$$\frac{1}{\pi}\arccos\frac{1}{3} = \frac{\varphi}{\pi} = -\frac{n_1 + \dfrac{n_2}{2}}{n_3}$$

is *rational*, contrary to Lemma 9. Therefore, $n_3 = 0$, and hence, using (20), we find that

$$n_1 f(\pi) + n_2 f\left(\frac{\pi}{2}\right) + n_3 f(\varphi) = 0.$$

But this means that the function f is additive.

Finally we calculate the Dehn invariants. Let l be the length of an edge of the cube Q. Then

$$f(Q) = 12\, l f\left(\frac{\pi}{2}\right) = 0$$

(see (20)), since Q has 12 edges. Moreover, letting m be the length of an edge of the regular tetrahedron P, we find that

$$f(P) = 6mf(\varphi) = 6m \neq 0.$$

Thus $f(P) \neq f(Q)$, and hence, by Theorem 19, the regular tetrahedron P and the cube Q are nonequidecomposable (and nonequicomplementable).[15]

Theorem 20 shows, as foreseen by Hilbert, that the concepts of equality of volume and equidecomposability (or equicomplementability) are *not equivalent* for polyhedra.

We now give further examples of the calculation of Dehn invariants. First of all, we prove that *the Dehn invariant of any parallelepiped is zero* (for any choice of additive function). In fact, let ab and cd be two opposite sides of a face $abcd$ of a parallelepiped P. If α is the dihedral angle at the edge ab, then $\pi - \alpha$ is the dihedral angle at the edge cd. Thus the set M on which the additive function f (used to calculate the Dehn invariant $f(P)$) is defined must contain the numbers π, α, $\pi - \alpha$ (and perhaps other numbers as well). Between these numbers there is the linear dependence

$$-\pi + \alpha + (\pi - \alpha) = 0,$$

and hence,

$$-f(\pi) + f(\alpha) + f(\pi - \alpha) = 0$$

by the additivity of the function f. Moreover $f(\pi) = 0$, since this condition is always imposed in considering Dehn invariants (cf. Theorem 19). Thus $f(\alpha) + f(\pi - \alpha) = 0$. Finally, since the edges ab and cd have the same length l, the terms $lf(\alpha)$ and $lf(\pi - \alpha)$ corresponding to these edges have a sum equal to zero, and the same is true of the terms corresponding to the other two

---

[15] As shown by Lebesgue [42], no two regular polyhedra (of equal volume) can be equidecomposable, i.e., their Dehn invariants are unequal (for a suitably chosen additive function f satisfying the condition $f(\pi) = 0$).

edges parallel to *ab* and *cd*. Thus the quadruple of parallel edges contributes zero to the sum defining the invariant $f(P)$, and the same is true of the other two quadruples of parallel edges. Therefore $f(P) = 0$, as asserted.

As a second example, we prove the following more general result: *The Dehn invariant of every prism equals zero*. In fact, let Q be an *n*-gonal prism, and let *abcd* be a lateral face of Q, where the edges *ab* and *cd*, both of length *l*, say, belong to the bases of the prism. Then the dihedral angles at these edges equal α and $\pi - \alpha$, respectively, and, as before, this means that the terms $lf(\alpha)$ and $lf(\pi - \alpha)$ corresponding to the edges *ab* and *cd* have a sum equal to zero. Thus the terms corresponding to all the edges of the upper and lower bases of the prism contribute zero to the sum defining the Dehn invariant $f(Q)$, but we must still consider the contributions of the *n* lateral edges of Q, each of length *m*, say. As a result, we get

$$f(Q) = mf(\alpha_1) + mf(\alpha_2) + \ldots + mf(\alpha_n),$$

where $\alpha_1, \alpha_2, \ldots, \alpha_n$ are the dihedral angles at the lateral edges. But $\alpha_1, \ldots, \alpha_n$ are the interior angles of the *n*-gon which is the perpendicular cross section of Q, and therefore

$$\alpha_1 + \alpha_2 + \ldots + \alpha_n - (n - 2)\pi = 0.$$

Because of the additivity of the function f, this implies

$$f(\alpha_1) + f(\alpha_2) + \ldots + f(\alpha_n) - (n - 2)f(\pi) = 0.$$

Bearing in mind that $f(\pi) = 0$, we find that $f(Q) = 0$, as asserted.

In the examples just considered, the Dehn invariants were calculated directly. However, the formula $f(Q) = 0$ for an arbitrary prism Q (and, in particular, for a parallelepiped) can be deduced from Theorem 19. In fact, since every prism is equidecomposable with some rectangular parallelepiped A, it follows from Theorem 19 that the formula $f(Q) \neq f(A)$ cannot hold, and hence $f(Q) = f(A) = 0$. Similarly, we can use Theorem 19 to calculate the Dehn invariant of Hill's tetrahedron. Since this tetrahedron (which we denote by H) is equidecomposable with some cube A, it follows from Theorem 19 that the formula $f(H) \neq f(A)$ cannot hold, and hence $f(H) = f(A) = 0$.

Next we give another example of a polyhedron whose Dehn invariant is nonzero (for a suitable choice of the additive function f). Consider the tetrahedron R whose edges cb, ac and cd are mutually perpendicular and have the same length (Fig. 61). If e is the midpoint of the edge ab, then the lengths of the segments ab, ce and de are $l\sqrt{2}$, $l/\sqrt{2}$ and $l\sqrt{3}/\sqrt{2}$, respectively. Therefore, letting α denote the dihedral angle at the edge ab, we find by examining the right triangle cde that $\cos \alpha = 1/\sqrt{3}$. Moreover, the dihedral angles at the edges bd and ad are also equal to α, while the dihedral angles at the edges cb, ac and cd are equal to $\pi/2$. On the set $M = \{\pi, \pi/2, \alpha\}$ we define a real function f, by setting

$$f(\pi) = 0, \quad f\left(\frac{\pi}{2}\right) = 0, \quad f(\alpha) = 1.$$

For every positive integer k we have $\cos k\alpha = a_k/(\sqrt{3})^k$, where a_k is an integer relatively prime to the number 3 (the proof of this fact is analogous to the proof of Lemma 9), from which it follows that the number $\dfrac{\alpha}{\pi} = \dfrac{1}{\pi}$ arc $\cos \dfrac{1}{\sqrt{3}}$ is *irrational.* The proof of the *additivity* of the function f under consideration is then literally the same as in the proof of Theorem 20. We now have

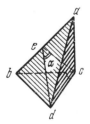

FIG. 61

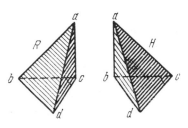

FIG. 62

$$f(R) = 3l \cdot f\left(\frac{\pi}{2}\right) + 3l\sqrt{2} \cdot f(\alpha) = 3l\sqrt{2} \neq 0.$$

Since $f(A) = 0$ for any cube A, it follows from Theorem 19 that the tetrahedron R and the cube of the same volume as R are nonequidecomposable.

Finally, we compare the tetrahedron R with Hill's tetrahedron H (Fig. 62). The tetrahedron H is equidecomposable with the cube of the same volume, but the tetrahedron R is not, as just shown. Therefore, R and H are *not* equidecomposable (and not equicomplementable) with each other. At the same time, these tetrahedra have congruent bases (the isosceles right triangle bcd, whose legs are of length l) and the same altitude l. This is the very example which Hilbert was so interested in finding (see the last sentence of his formulation of the problem).

§14. Hadwiger's theorem

To validate the results expounded in the preceding section, we must still prove Theorem 19, and the present section is devoted to this task. Hadwiger's original form of the theorem will be given at the end of the section (Theorem 21).

Lemma 10. *Let f be an additive function defined on a set M, let* γ *be a real number which does not belong to M, and let M* be the set obtained from M by adjunction of the element* γ, *so that M** =

$M \cup \{\gamma\}$. *Then the function f can be extended onto the set M^*, i.e., there exists an additive function which is defined on the set M^* and coincides with f on the set M.*

PROOF. We consider two cases.

Case 1. There is no linear dependence

$$n_1 x_1 + n_2 x_2 + \ldots + n_k x_k + n\gamma = 0,$$

between the elements of the set M^* in which the coefficient of the number γ is nonzero. In other words, the number γ does not appear in any linear dependence. In this case, the number $f(\gamma)$ is not restricted by any conditions, i.e., we can choose $f(\gamma)$ to be any real number at all.

Case 2. There is a linear dependence

$$n_1 x_1 + n_2 x_2 + \ldots + n_k x_k + n\gamma = 0, \quad n \neq 0, \quad (22)$$

involving the number γ. In this case, we fix *one* linear dependence of the form (22) and use it to define the number $f(\gamma)$ from the formula

$$n_1 f(x_1) + n_2 f(x_2) + \ldots + n_k f(x_k) + n f(\gamma) = 0, \quad (23)$$

i.e., we set

$$f(\gamma) = -\frac{n_1}{n} f(x_1) - \frac{n_2}{n} f(x_2) - \ldots - \frac{n_k}{n} f(x_k).$$

As we now show, this method leads to an additive function on the set M^*. In fact, consider an arbitrary l. d. (linear dependence) between the elements of the set M^*. If the number γ does not appear in this l.d., we agree to regard γ as appearing in the l. d. with coefficient 0. In the same way, we can assume that all the numbers $x_1, \ldots, x_k$ appear in the l. d. (possibly with zero coefficients). Finally, certain numbers $y_1, \ldots, y_l \in M$, different from $x_1, \ldots, x_k$, may appear in the l. d. Thus the given l. d. can be written in the form

$$m_1 x_1 + \ldots + m_k x_k + p_1 y_1 + \ldots + p_l y_l + m\gamma = 0, \tag{24}$$

and we must prove that there is an l. d. with the same coefficients between the corresponding values of f, i.e., that

$$m_1 f(x_1) + \ldots + m_k f(x_k) + p_1 f(y_1) + \ldots + p_l f(y_l) + \\ + mf(\gamma) = 0. \tag{25}$$

To this end, we multiply equation (24) by n and then subtract the result from equation (22) multiplied by m, obtaining

$$(m_1 n - mn_1) x_1 + \ldots + (m_k n - mn_k) x_k + \\ + p_1 n y_1 + \ldots + p_l n y_l = 0. \tag{26}$$

This is an l. d. between elements of the set M, and hence

$$(m_1 n - mn_1) f(x_1) + \ldots + (m_k n - mn_k) f(x_k) + \\ + p_1 n f(y_1) + \ldots + p_l n f(y_l) = 0$$

(since the function f is additive on the set M). Adding equation (23) multiplied by m to this equation, we obtain

$$m_1 n f(x_1) + \ldots + m_k n f(x_k) + \\ + p_1 n f(y_1) + \ldots + p_l n f(y_l) + mnf(\gamma) = 0.$$

Finally, dividing the last equation by the number $n \neq 0$, we get the desired result (25). Thus the function f is also additive on the set M^*.

   Lemma 11. *Let* A, $P_1, \ldots, P_k$ *be polyhedra such that* $A = P_1 + \ldots + P_k$, *and let* M *be the the set containing* π *and all the dihedral angles of all the polyhedra* $A, P_1, \ldots, P_k$. *Moreover, let* f *be an additive function which is defined on the set* M *and satisfies the condition* $f(\pi) = 0$. *Then the Dehn invariants of the given polyhedra are related by the formula*

$$f(A) = f(P_1) + \ldots + f(P_k). \qquad (27)$$

PROOF. We consider all the line segments that are edges of the polyhedra A, P_1, ..., P_k, and on these segments we mark all the points that are vertices of A, P_1, ..., P_k, together with all the points in which the edges intersect each other. This gives a finite number of segments (in general, shorter than the edges) which we call *links*, following V. F. Kagan. For example, Fig. 63 shows a decomposition of a cube into polyhedra (cf. Fig. 64), in which the edge l_1 consists of three links m_1, m_2 and m_3. In general, each edge of each polyhedron consist of one or several links.

Given any link lying on an edge of the polyhedron A, let m be the length of the link and let α be the corresponding dihedral angle of A. Then $\alpha \in M$, and hence the number $f(\alpha)$ is defined. We call the product $mf(\alpha)$ the *weight* of the given link in the polyhedron A, and the weights of the links in the polyhedra P_1, ..., P_k are defined in the same way.

We now take all the links lying on the edges of the polyhedron A, find their weights in A, and calculate the sum of all these weights. It is not hard to see that this sum equals the Dehn invariant $f(A)$. For example, if the edge l_1 of the polyhedron A consists of three links, of lengths m_1, m_2, m_3 (see Fig. 63), then the edge l_1 and the links m_1, m_2, m_3 are associated with the same dihedral angle α_1 in the

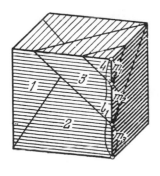
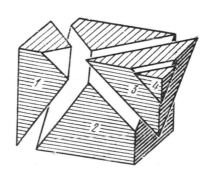

FIG. 63   FIG. 64

polyhedron A, and hence the sum of the weights of the links $m_1, m_2,$ m_3 equals

$$m_1 f\,(\alpha_1) + m_2 f\,(\alpha_1) + m_3 f\,(\alpha_1) = (m_1 + m_2 + m_3) f\,(\alpha_1)$$
$$= l_1 f\,(\alpha_1).$$

In just the same way, the sum of the weights of all the links making up the edge l_2 of the polyhedron A equals $l_2 f\,(\alpha_2)$, where α_2 is the dihedral angle at the edge l_2; and so on. Hence the sum of the weights of all the links lying on the edges of the polyhedron A equals the Dehn invariant of A, and similarly, the Dehn invariant of each polyhedron P_i equals the sum of the weights (in P_i) of all the links lying on the edges of P_i.

To calculate the sum in the right-hand side of equation (27), we must form the sum of the weights of all the links in all the polyhedra $P_1, \ldots, P_k$. Let us find the coefficient with which a given link m enters into this sum. If the dihedral angles of the polyhedra $P_1, \ldots,$ P_k which adjoin the link m are denoted by $\gamma_1, \ldots, \gamma_s$ (where these numbers belong to the set M), then the weight of the link m in the polyhedron with dihedral angle γ_1 equals $mf\,(\gamma_1)$, its weight in the polyhedron with dihedral angle γ_2 equals $mf\,(\gamma_2)$, and so on. Thus the sum of the weights of the link m in all the polyhedra $P_1, \ldots, P_k$ adjoining the link is equal to

$$mf\,(\gamma_1) + mf\,(\gamma_2) + \ldots + mf\,(\gamma_s). \tag{28}$$

To continue our analysis, we divide the set of all links into three groups.

1. Links which lie entirely inside the polyhedron A (with the possible exception of their end points). If m is such a link and if each of the polyhedra $P_1, \ldots, P_k$ adjoining this link has the link on one of its *edges*, then the sum of the dihedral angles adjoining the link m is a full $360°$ (see Fig. 65, where in this figure, as in the ones that follow, we show the intersection of the polyhedron A and of the polyhedra adjoining the link m with the plane perpendicular to m, the link m itself being indicated by a single point r). Therefore $\gamma_1 +$ $\ldots + \gamma_s = 2\pi$ in this case, i.e.,

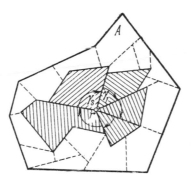

FIG. 65

$$\gamma_1 + \ldots + \gamma_s - 2\pi = 0.$$

But this is a linear dependence between elements of the set M, and hence, by the additivity of the function f, we have

$$f(\gamma_1) + \ldots + f(\gamma_s) - 2f(\pi) = 0,$$

i.e., $f(\gamma_1) + \ldots + f(\gamma_s) = 0$, since $f(\pi) = 0$, by hypothesis. Thus the expression (28) vanishes in this case.

On the other hand, if m is a link lying inside the polyhedron A, but if one[16] of the polyhedra $P_1, \ldots, P_k$ adjoining the link has m on one of its *faces*, then the dihedral angles of the remaining polyhedra form an angle of 180° (Fig. 66), i.e., $\gamma_1 + \ldots + \gamma_s = \pi$. This implies, as before, that the expression (28) vanishes. Thus links which lie *inside* the polyhedron A can be disregarded in calculating the right-hand side of equation (27), since the sum of the weights of such a link equals zero.

[16] If *two* polyhedra adjoining a segment m do not have m on their edges, i.e., if the segment m lies on the *faces* of two polyhedra that adjoin each other, then these two polyhedra are the only ones that adjoin m. Then the segment m cannot lie on the edges of any of the polyhedra $P_1, \ldots, P_k$, so that m cannot be a link.

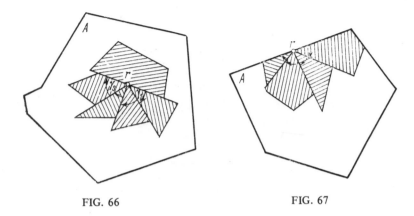

FIG. 66　　　　　　　　FIG. 67

2. Links which lie on the faces of the polyhedron A, but not on its edges. In this case $\gamma_1 + \ldots + \gamma_s = \pi$ (Fig. 67), and the expression (28) vanishes, just as in the preceding case.

3. We must still consider links lying on the edges of the poly-hedron A. In this case the sum $\gamma_1 + \ldots + \gamma_s$ equals α or $\alpha - \pi$, where α is the dihedral angle of the polyhedron A (Fig. 68). In both

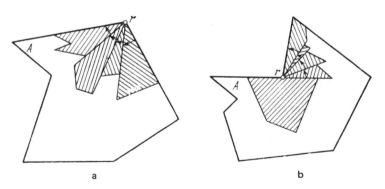

a　　　　　　　　　　b

FIG. 68

cases we have

$$f(\gamma_1) + \ldots + f(\gamma_s) = f(\alpha),$$

and the expression (28) is equal to $mf(\alpha)$, i.e., to the weight of the given link in the polyhedron A. Thus, finally, the sum in the right-hand side of (27) is equal to the sum of the weights of all the links of the polyhedron A, i.e., to the Dehn invariant $f(A)$.

Proof of Theorem 19. Let A and B be equidecomposable, so that

$$A = P_1 + \ldots + P_k, \quad B = Q_1 + \ldots + Q_k;$$
$$P_1 \cong Q_1, \ldots, P_k \cong Q_k.$$

and let f be an additive function defined on the set M figuring in the statement of Theorem 19. Then, according to Lemma 10, by adding new numbers one after another to the set M, we can extend f onto the set M' containing all the elements of M and all the dihedral angles of all the polyhedra $P_1, \ldots, P_k$ (the congruent polyhedra $Q_1, \ldots, Q_k$ have the same dihedral angles). By Lemma 11, we then have

$$f(A) = f(P_1) + \ldots + f(P_k),$$
$$f(B) = f(Q_1) + \ldots + f(Q_k).$$

Moreover, since the polyhedra P_i and Q_i are congruent, the corresponding edge lengths and dihedral angles of P_i and Q_i are equal, so that $f(P_i) = f(Q_i)$, $i = 1, \ldots, k$. Therefore $f(A) = f(B)$, and the theorem is proved.

Theorem 19 can be given another form by using the concepts introduced in §4, and this is the form in which the theory of equidecomposability appears in the work of Hadwiger ([21], [24], [26], [27], etc.).

Theorem 21. *Let f be an additive function which is defined on the whole number line R and satisfies the condition $f(\pi) = 0$. Then the Dehn invariant $f(A)$ is an additive D-invariant, where D is the group of all motions of the space R^3.*

PROOF. Since the function f is defined on the set of all real numbers, the Dehn invariant $f(A)$ is defined on the set of all polyhedra. If $A = P_1 + \ldots + P_k$, then $f(A) = f(P_1) + \ldots + f(P_k)$ (Lemma 11), i.e., the function $f(A)$ is *additive*. Finally, the D-invariance of $f(A)$ is obvious, since if $A \cong B$, then the corresponding edge lengths and dihedral angles of the polyhedra A and B are equal, and therefore $f(A) = f(B)$.

Corollary. *Let f have the same meaning as in Theorem 21. Then a necessary condition for equidecomposability (and also for equicomplementability) of two polyhedra A and B is that $f(A) = f(B)$.*

This is an immediate consequence of the preceding theorem (by Theorem 15, which, together with its proof, continues to hold for polyhedra).

Next we show how Theorem 20 can be deduced from this corollary. Let P be the regular tetrahedron and Q the cube with the same volume, and let f be any Cauchy–Hamel function which satisfies the conditions $f(\pi) = 0$, $f(\varphi) = 1$, where $\varphi = \text{arc cos } 1/3$ is the dihedral angle of P.[17] Then $f(\pi/2) = 0$, and hence $f(Q) = 0$. At the same time, $f(P) = 6mf(\varphi) \neq 0$ (where m is the length of an edge of P). Thus $f(P) \neq f(Q)$ and hence, by the above corollary, the polyhedra P and Q are nonequidecomposable (and nonequicomplementable).

This proof, using Hadwiger's results (Theorem 21 and its corollary) is conceptually somewhat "purer" than the proof of Theorem 20 given in the preceding section. This conceptual purity consists, for example, in the fact that here we consider *additive D-invariants* defined at once on the set of all polyhedra, instead of the Dehn invariants which in §13 were defined only for certain polyhedra (namely those whose dihedral angles belong to the set M on which f is given). This fact is of no small importance, since it will allow us to

[17] Since the ratio φ/π is irrational (Lemma 9), there exists a *rational basis* for the real line R which contains the numbers π and φ (see p. 29). We set $f(\pi) = 0$, $f(\varphi) = 1$ and define the values of the function f arbitrarily on the remaining elements of the chosen basis. At the other points of the number line (which are rational linear combinations of the basis elements) the values of f are uniquely defined by formula (6), p. 28. This gives the required function f.

regard the Dehn invariants as homomorphisms of the algebra of polyhedra (see §22). At the same time, Hadwiger's proof of Theorem 20, which uses Cauchy-Hamel functions, rests in a fundamental way on the axiom of choice, while our "finite" version of the proof, based on the use of Theorem 19, is free from the use of the axiom of choice (and hence is more elementary). Thus the two versions of the proof differ only in their algebraic form, since they have the same "geometric kernel" (in our treatment), consisting of the use of Lemma 11.

In conclusion, we note that the solution of Hilbert's third problem can be presented somewhat differently, in the spirit of Hadwiger's ideas (without focusing our attention on the nonequidecomposability of the regular tetrahedron and the cube, or on the existence of two nonequidecomposable tetrahedra with congruent bases and the same altitude). Consider the same Cauchy-Hamel function $f(x)$ as above. According to Theorem 21, the corresponding Dehn invariant $f(A)$, defined on the set of all polyhedra, is an additive D-invariant. Moreover, $f(P) = 0$ for every parallelepiped (or prism) P and $f(B) \neq 0$ if B is a regular tetrahedron. Now let

$$v^*(A) = v(A) + f(A),$$

where v is the volume. The function $v^*(A)$, which is defined on the set of all polyhedra, is an additive D-invariant, and $v^*(P) = v(P)$ for every parallelepiped (or prism) P, while $v^*(B) \neq v(B)$ for a regular tetrahedron B.

Thus $v^*(A)$ is an additive D-invariant, *different* from $v(A)$, which satisfies the condition

$$v^*(P) = v(P) = abc,$$

if P is the rectangular parallelepiped whose edges are of length a, b, c. In other words, axioms (β) and (γ^*), together with the formula for the volume of a rectangular parallelepiped, *do not permit* the construction of a unique function $v(A)$ (the volume) on the set of all polyhedra. And since the formula for the volume of a triangular pyramid, together with axioms (β) and (γ^*), *uniquely* determines the volume of an arbitrary polyhedron (cf. Figs. 56-57), *the formula for the volume of a triangular pyramid cannot be deduced from the formula for the volume of a rectangular parallelepiped, together with axioms* (β) *and* (γ^*). In other words, to derive the formula for the volume of a triangular pyramid, it is *necessary* to use the "non-elementary" axiom (α). In particular, the theorem on the equality of the volumes of two triangular pyramids with congruent bases and the same altitude *cannot be deduced* from the formula for the volume of a rectangular parallelepiped, together with axioms (β) and (γ^*) (since this is the only place in the theory of the volume of polyhedra where axiom (α) is used).

§15. Bricard's condition

In 1896 Bricard published a paper [9] containing the following assertion:

If two polyhedra A and B are equidecomposable, then integers n_i $> 0, n'_j > 0$ *and p can be found such that*

$$n_1\alpha_1 + \ldots + n_q\alpha_q = n'_1\beta_1 + \ldots + n'_r\beta_r + p\pi, \quad (29)$$

where $\alpha_1, \ldots, \alpha_q$ are the dihedral angles of the polyhedron A and $\beta_1, \ldots, \beta_r$ are the dihedral angles of the polyhedron B.

We will call this assertion *Bricard's condition*. In particular, if B is a cube, then the right-hand side of (29) is of the form $p' \cdot \pi/2$, where p' is an integer (moreover $p' > 0$, since the left-hand side of (29) is positive). Thus we have (multiplying (29) by 2, if necessary) the following special case of Bricard's condition:

If a polyhedron A is equidecomposable with a cube, then positive integers $n_1, \ldots, n_q$, p can be found such that

$$n_1\alpha_1 + \ldots + n_q\alpha_q = p\pi, \quad (30)$$

where $\alpha_1, \ldots, \alpha_q$ are the dihedral angles of the polyhedron A.

Formula (30) immediately implies Theorem 20 (which was also stated in reference [9]). Thus Bricard's paper, which appeared four years before Hilbert posed the problem, already contained the formulation of the theorem on the nonequidecomposability of the regular tetrahedron and the cube! However, the proof given by Bricard was incorrect (and presumably this is why Hilbert did not cite Bricard's paper).

Let us now follow Bricard's train of thought. Let the polyhedra A and B be equidecomposable, so that

$$A = P_1 + \ldots + P_k, \quad B = Q_1 + \ldots + Q_k;$$
$$Q_i = f_i(P_i), \quad i = 1, \ldots, k,$$

where $f_1, \ldots, f_k$ are certain motions. As in the proof of Lemma 11, we divide the edges of all the polyhedra into *links*. Suppose that we have succeeded (by making the links shorter, if necessary) in finding a decomposition into links which is invariant under the motions f_i (i.e., such that f_i carries the links of the polyhedron P_i into the links of the polyhedron Q_i). Let Σ denote the sum of the dihedral angles of the polyhedra $P_1, \ldots, P_k$ over all the links (so that if an edge with dihedral angle α in the polyhedron P_i is divided into five links, say, then the term 5α appears in the sum Σ). If a link m lies inside the polyhedron A or on one of its faces, then, as we saw in the proof of Lemma 11, the sum of the dihedral angles of the polyhedra P_i adjoining the link equals 2π or π. On the other hand, if a link m lies on an edge l of the polyhedron A, then the sum of the dihedral angles of the polyhedra P_i adjoining the link equals α or $\alpha - \pi$, where α is the dihedral angle of the polyhedron A at the link m (i.e., at the edge l). Therefore,

$$\Sigma = n\pi + n_1\alpha_1 + \ldots + n_q\alpha_q,$$

where $n_1 > 0, \ldots, n_q > 0$ and n are integers (more exactly, n_i is the number of links making up the edge whose dihedral angle is α_i, since the term α_i or $\alpha_i - \pi$ is obtained from every link lying along this edge). Moreover, since the mapping f_i carries the links of the polyhedron P_i into the links of the polyhedron Q_i, the sum over all the links of the dihedral angles of the polyhedra $Q_1, \ldots, Q_k$ is also equal to Σ. Therefore, as before,

$$\Sigma = n'\pi + n_1'\beta_1 + \ldots + n_r'\beta_r,$$

where $n_1' > 0, \ldots, n_r' > 0$ and n' are integers, and equating the two values for Σ just found, we get formula (29).

Thus the above argument contains a proof of formula (29), provided that there exists an invariant decomposition into links. Bricard erroneously assumed that such a decomposition always exists, and this, as we now show, is *false*.

Consider the two equidecomposable polyhedra $A = P_1 + P_2 + P_3$ and $B = Q_1 + Q_2 + Q_3$ shown in Fig. 69. Here $Q_i = f_i(P_i)$, $i = 1, 2, 3$, where f_1, f_2 are translations, and f_3 is a twist (the composition of a rotation through the angle $\pi/2$ about the line ab and a shift along ab). The polyhedra P_1, P_2, P_3 have the same edge ab, which is carried by the motions f_1, f_2, f_3 into the segments pq, mn, qr, respectively. We will assume that the length of the edge ab equals 1, while the length of the segment pq (which we denote by ξ) is an irrational number less than 1 (for example, $\xi = \sqrt{2} - 1$). Suppose there exists an invariant decomposition into links, and let W denote the set of all end points of the links. Then the set W is *finite*.

Now let $F(x)$ denote the *fractional part* of the number x (i.e., $0 \leqslant F(x) < 1$ and $x - F(x)$ is an integer). It is not hard to show

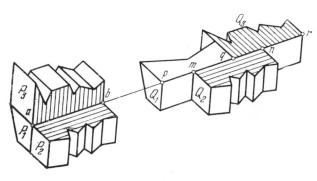

FIG. 69

that for every positive integer k the point m_k of the segment pq lying at the distance $F(k\xi)$ from p belongs to the set W. In fact, for $k = 1$ we have $F(k\zeta) = F(\xi) = \xi$, i.e., the point m_1 coincides with m and hence belongs to W. Suppose it has already been shown that $m_k \in W$. Then $h_k = f_1^{-1}(m_k) \in W$ and hence $n_k = f_2(h_k) \in W$. The point n_k lies at a distance $F(k\xi) + \xi$ from p. If $F(k\xi) + \xi < 1$, then $F(k\xi) + \xi = F((k+1)\xi)$, i.e., n_k coincides with m_{k+1}, and hence $m_{k+1} \in W$. However, if $F(k\xi) + \xi \geqslant 1$, then the point n_k lies on the segment qr, and its distance from the point q is $F(k\xi) + \xi - 1 = F((k+1)\xi)$. But then the point $f_1(f_3^{-1}(n_k))$, belonging to W, lies at a distance $F((k+1)\xi)$ from p, i.e., coincides with m_{k+1}. Therefore, $m_{k+1} \in W$ in this case too.

Thus all the points m_k, $k = 1, 2, \ldots$ belong to W. But, since the number ξ is irrational, all the numbers $F(k\xi)$, $k = 1, 2, \ldots$ are *different*, i.e., all the points m_k, $k = 1, 2, \ldots$ are different, and hence the set W must be *infinite*. This contradiction shows that an invariant decomposition into links (with respect to the motions f_1, f_2, f_3) does not exist. Thus the proof proposed by Bricard contains a gap that cannot be filled. Nevertheless, it can be shown that Bricard's condition is valid in any event (as proved by Dehn himself [12]).

To this end, let

$$M = \{\alpha_1, \ldots, \alpha_q, \beta_1, \ldots, \beta_r, \pi\}.$$

Every linear dependence between the elements of M is of the form

$$r_1\alpha_1 + \ldots + r_q\alpha_q + s_1\beta_1 + \ldots + s_r\beta_r = t\pi,$$

where the r_i, s_j, t are rational numbers (in fact, integers), not all equal to zero. This linear dependence can be written as the equation

$$r_1x_1 + \ldots + r_qx_q + s_1y_1 + \ldots + s_ry_r = t\pi,$$

with the solution $x_i = \alpha_i$, $y_j = \beta_j$. The system of *all* such equations is compatible, since it has the solution $x_i = \alpha_i$, $y_j = \beta_j$. It follows from the theory of linear equations that this system (consisting of infinitely many equations) can be solved for certain unknowns, i.e., it is equivalent to a system which (after renumbering the unknowns, if necessary) can be written in the form

$$
\begin{cases}
x_i = \sum_{j=\lambda+1}^{q} a_{ij}x_j + \sum_{j=\mu+1}^{r} b_{ij}y_j + c_i\pi, & i = 1, \ldots, \lambda; \\
y_i = \sum_{j=\lambda+1}^{q} a'_{ij}x_j + \sum_{j=\mu+1}^{r} b'_{ij}y_j + c'_i\pi, & i = 1, \ldots, \mu,
\end{cases}
$$

where the a_{ij}, b_{ij}, c_i, a'_{ij}, b'_{ij}, c'_i are certain rational numbers, and $x_{\lambda+1}, \ldots, x_q, y_{\mu+1}, \ldots, y_r$ are the "free" unknowns. In other words, the elements of the set M satisfy the linear relations

$$
\begin{cases}
\alpha_i = \sum_{j=\lambda+1}^{q} a_{ij}\alpha_j + \sum_{j=\mu+1}^{r} b_{ij}\beta_j + c_i\pi, & i = 1, \ldots, \lambda; \\
\beta_i = \sum_{j=\lambda+1}^{q} a'_{ij}\alpha_j + \sum_{j=\mu+1}^{r} b'_{ij}\beta_j + c'_i\pi, & i = 1, \ldots, \mu,
\end{cases}
\tag{31}
$$

and all the remaining linear relations between elements of M are a consequence of these. It follows that a function f defined on M is additive if and only if the following formulas hold:

$$
\begin{cases}
f(\alpha_i) = \sum_{j=\lambda+1}^{q} a_{ij}f(\alpha_j) + \sum_{j=\mu+1}^{r} b_{ij}f(\beta_j) + c_if(\pi), \\
\qquad\qquad\qquad\qquad\qquad\qquad i = 1, \ldots, \lambda; \\
f(\beta_i) = \sum_{j=\lambda+1}^{q} a'_{ij}f(\alpha_j) + \sum_{j=\mu+1}^{r} b'_{ij}f(\beta_j) + c'_if(\pi), \\
\qquad\qquad\qquad\qquad\qquad\qquad i = 1, \ldots, \mu.
\end{cases}
\tag{32}
$$

Suppose now that $\lambda < q$. Fixing a number $\nu = \lambda + 1, \ldots, q$, we set

$$
f(\alpha_\nu) = 1,
$$

$f(\alpha_j) = 0$ for indices $j = \lambda + 1, \ldots, q$ other than ν,

$$f(\beta_j) = 0 \text{ for } j = \mu + 1, \ldots, r.$$

Using (32) to define the remaining numbers $f(\alpha_i)$ and $f(\beta_i)$, i.e.,

$$f(\alpha_i) = a_{i\nu}, \quad i = 1, \ldots, \lambda; \quad f(\beta_i) = a'_{i\nu}, \quad i = 1, \ldots, \mu,$$

we get an additive function f, which is defined on the set M and satisfies the condition $f(\pi) = 0$. The Dehn invariants $f(A)$ and $f(B)$ are then given by

$$f(A) = l_1 f(\alpha_1) + \ldots + l_q f(\alpha_q) = l_1 a_{1\nu} + \ldots + l_\lambda a_{\lambda\nu} + l_\nu,$$
$$f(B) = l'_1 f(\beta_1) + \ldots + l'_r f(\beta_r) = l'_1 a'_{1\nu} + \ldots + l'_\mu a'_{\mu\nu},$$

where $l_1, \ldots, l_q$ are the lengths of the edges of the polyhedron A and $l'_1, \ldots, l'_r$ those of the polyhedron B. But $A \sim B$ implies $f(A) = f(B)$, by Theorem 19. Therefore

$$\sum_{i=1}^{\lambda} l_i a_{i\nu} - \sum_{i=1}^{\mu} l'_i a'_{i\nu} = -l_\nu < 0,$$

where this formula holds for arbitrary $\nu = \lambda + 1, \ldots, q$, and similarly

$$-\sum_{i=1}^{\lambda} l_i b_{i\nu} + \sum_{i=1}^{\mu} l'_i b'_{i\nu} < 0, \quad \nu = \mu + 1, \ldots, r.$$

We now choose positive rational numbers $m_1, \ldots, m_\lambda$, $m'_1, \ldots, m'_\mu$ which are close enough to $l_1, \ldots, l_\lambda, l'_1, \ldots, l'_\mu$ to make all the numbers

$$\left\{ \begin{array}{ll} \displaystyle\sum_{i=1}^{\lambda} m_i a_{i\nu} - \sum_{i=1}^{\mu} m'_i a'_{i\nu} = -m_\nu, & \nu = \lambda + 1, \ldots, q; \\[4mm] \displaystyle -\sum_{i=1}^{\lambda} m_i b_{i\nu} + \sum_{i=1}^{\mu} m'_i b'_{i\nu} = -m'_\nu, & \nu = \mu + 1, \ldots, r \end{array} \right. \tag{33}$$

negative. Then *all* the numbers $m_1, \ldots, m_q, m_1', \ldots, m_r'$ will be positive and rational. It follows from (31) and (33) that

$$
\begin{aligned}
\sum_{i=1}^{q} m_i \alpha_i - \sum_{i=1}^{r} m_i' \beta_i & = \\
& = \sum_{i=1}^{\lambda} m_i \left(\sum_{j=\lambda+1}^{q} a_{ij}\alpha_j + \sum_{j=\mu+1}^{r} b_{ij}\beta_j + c_i\pi \right) - \\
& \quad - \sum_{v=\lambda+1}^{q} \alpha_v \left(\sum_{i=1}^{\lambda} m_i a_{iv} - \sum_{i=1}^{\mu} m_i' a_{iv}' \right) - \\
& \quad - \sum_{i=1}^{\mu} m_i' \left(\sum_{j=\lambda+1}^{q} a_{ij}'\alpha_j + \sum_{j=\mu+1}^{r} b_{ij}'\beta_j + c_i'\pi \right) - \\
& \quad - \sum_{v=\mu+1}^{r} \beta_v \left(\sum_{i=1}^{\lambda} m_i b_{iv} - \sum_{i=1}^{\mu} m_i' b_{iv}' \right) = \left(\sum_{i=1}^{\lambda} m_i c_i - \sum_{i=1}^{\mu} m_i' c_i' \right) \pi,
\end{aligned}
$$

where the numbers m_i, m_i' are positive and rational, while the numbers c_i, c_i' are rational. Therefore, getting rid of denominators, we finally arrive at a relation of the form (29). The proof of Bricard's condition is now complete. It is clear from the proof that Bricard's condition is a consequence of Theorem 19, i.e., of the *Dehn* conditions for equidecomposability.

§16. Equivalence of the methods of decomposition and complementation

In §7 we saw that the methods of decomposition and complementation are equivalent for polygons (in the Euclidean plane), i.e., in this case equidecomposability and equicomplementability mean exactly the same thing. The proof of this fact consisted of two parts. The fact that equidecomposability of two polygons implies their equicomplementability (Theorem 11) remains valid in different G-geometries, as noted on p. 77. This property, together with its proof, also remains valid for *polyhedra* in three-dimensional (or *n*-dimensional) space. The proof of the converse assertion (that equicomplementability of two polygons implies their equidecomposability), as given in §7, makes essential use of the Bolyai–Gerwien theorem on the equidecomposability of *any* two polygons of equal area. Thus, if the analogue of the Bolyai–Gerwien theorem (i.e., the equivalence of equality of area and equidecomposability) fails to hold in some geometry, then the problem of the equivalence of the methods of decomposition and complementation cannot be solved as easily as in §7, but requires further investigation.[18]

For example, in non-Archimedean geometry (§8), where there is no analogue of the Bolyai–Gerwien theorem, the methods of decomposition and complementation are not equivalent (the triangles in Fig. 33 have the same area and hence are equicomplementable, by Theorem 10_C, but they are not equidecomposable). A somewhat similar situation occurs for polyhedra in three-dimensional space. In fact, the solution of Hilbert's third problem given above shows that equality of volume and equidecomposability are *nonequivalent* concepts for polyhedra, but, unlike the case of non-Archimedean geometry, equidecomposability and equicomplementability are *equivalent* properties for polyhedra, as

---

[18] In reference [25] Hadwiger considers an arbitrary *crystallographic group* G, i.e., a discrete group of motions of the space R^n with a bounded *fundamental domain*. The equivalence of the methods of decomposition and complementation can also be proved under these conditions, but the form of the argument is somewhat different. Namely, Hadwiger constructs a system of additive G-invariants φ_a and shows that a *necessary and sufficient* condition for G-equidecomposability of the polyhedra A and B is that $\varphi_a(A) = \varphi_a(B)$ for all the invariants φ_a of the system in question. If now the polyhedra A and B are G-equicomplementable, then, by Theorem 15, $\varphi(A) = \varphi(B)$ for *every* additive G-invariant φ, and hence $A \underset{G}{\sim} B$, by what was just said.

The Hadwiger invariants φ_a for the crystallographic group G are constructed as follows: Given an arbitrary polyhedron A and a point $a \in R^n$, let $v_a(A)$ denote the ratio $v(A \cap S)/v(S)$, where v is the ordinary volume and S is a sufficiently small ball with center a (this ratio is independent of the radius r of the ball, if r is sufficiently small). Then

$$\varphi_a(A) = \sum_{g \in G} v_{g(a)}(A),$$

where only a finite number of terms in the sum are nonzero.

proved by Sydler [50]. This and related topics will be considered in the present section.

The original proof, due to Sydler, was generalized by Hadwiger to the case of n-dimensional Euclidean space [22], [28]. A completely different proof was subsequently found by Zylev [59], which is more formal, but has the advantages of brevity and greater generality. This is the proof that we give here (in a somewhat modified form).

Lemma 12. *Let A and B be two polyhedra in R^n satisfying the condition $v(A) > v(B)$. Then there exists a decomposition $A = M + N$ such that $M \underset{T}{\sim} B$.*

PROOF. Let $\varepsilon = \frac{1}{2}(v(A) - v(B))$. Choosing some ($n$-dimensional) mosaic, we let Q denote the union of all the cubes of the mosaic which have points in common with the polyhedron B, while P denotes the union of all the cubes of the mosaic contained in A. By refining the mosaic sufficiently, we can see to it that the inequalities $v(Q) - v(B) < \varepsilon$, $v(A) - v(P) < \varepsilon$ hold. Therefore, $v(Q) < v(P)$, i.e., the polyhedron P is made up of more cubes of the refined mosaic than Q. Let $K_1, \ldots, K_t$ be the cubes making up Q, so that $Q = K_1 + \ldots + K_t$, and let $L_1, \ldots, L_t$ be any distinct cubes of the mosaic for which $L_1 + \ldots + L_t \subset P \subset A$. Finally, let f_i be the translation carrying the cube K_i into L_i. Setting $U_i = B \cap K_i$, $V_i = f_i(U_i)$, $M = V_1 + \ldots + V_t$, we find that

$$B = B \cap (K_1 + \ldots + K_t) = U_1 + \ldots + U_t \underset{T}{\sim}$$

$$\underset{T}{\sim} V_1 + \ldots + V_t = M \subset A.$$

To finish the proof, we choose N to be the complement of the polyhedron M relative to A, i.e., $N = \overline{A \setminus M}$.

Theorem 22 (see [22]). *Let G be a group of motions of the space R^n which contains all translations. Then two polyhedra in R^n are G-equicomplementable if and only if they are G-equidecomposable.*

The proof (given below) is based on the ideas of reference [59]. First we need the following

Lemma 13. *Suppose that*

$$A + P_1 + \ldots + P_k \cong_G B + Q_1 + \ldots + Q_k,$$

where $P_i \underset{G}{\sim} Q_i,$ $i = 1, \ldots, k,$ *and* $2v(P_k) < v(A).$ *Then there exist polyhedra* $A^{(1)}, P_1^{(1)}, \ldots, P_{k-1}^{(1)},$ *such that*

$$A^{(1)} + P_1^{(1)} + \ldots + P_{k-1}^{(1)} \cong_G B + Q_1 + \ldots + Q_{k-1}, \quad (34)$$

$$A^{(1)} \underset{G}{\sim} A, \; P_1^{(1)} \underset{G}{\sim} Q_1, \ldots, P_{k-1}^{(1)} \underset{G}{\sim} Q_{k-1}. \quad (35)$$

PROOF. Let $f \in G$ be the motion such that

$$f(B + Q_1 + \ldots + Q_k) = A + P_1 + \ldots + P_k \quad (36)$$

and let $f(Q_k) \cap P_i = U_i,$ $V_i = \overline{P_i \setminus U_i},$ so that

$$P_i = U_i + V_i, \quad i = 1, \ldots, k. \quad (37)$$

Then

$$v(U_1 + \ldots + U_k) = v(f(Q_k) \cap (P_1 + \ldots + P_k)) \leqslant$$
$$\leqslant v(f(Q_k)) = v(Q_k) = v(P_k) < v(A) - v(P_k) =$$
$$= v(A) - v(f(Q_k)) \leqslant v(\overline{A \setminus f(Q_k)}).$$

Hence, by Lemma 12, there exist (since $T \subset G$) polyhedra $W_1 \underset{G}{\sim} U_1, \ldots, W_k \underset{G}{\sim} U_k$ and N such that

$$\overline{A \setminus f(Q_k)} = N + W_1 + \ldots + W_k,$$

and therefore

$$A = (f(Q_k) \cap A) + \overline{A \setminus f(Q_k)} =$$
$$= (f(Q_k) \cap A) + N + W_1 + \ldots + W_k. \quad (38)$$

We have

$$f(Q_k) = f(Q_k) \cap (A + P_1 + \ldots + P_k) =$$
$$= (f(Q_k) \cap A) + (f(Q_k) \cap P_1) + \ldots + (f(Q_k) \cap P_k) =$$
$$= (f(Q_k) \cap A) + U_1 + \ldots + U_k,$$

and hence, by (37) and (38),

$$A + P_1 + \ldots + P_k = (f(Q_k) \cap A) + N + W_1 + \ldots$$
$$\ldots + W_k + U_1 + \ldots + U_k + V_1 + \ldots + V_k =$$
$$= f(Q_k) + N + V_1 + \ldots + V_k + W_1 + \ldots + W_k.$$

Thus, setting

$$A^{(1)} = N + V_k + W_k, \ P_1^{(1)} = V_1 + W_1, \ \ldots, \ P_{k-1}^{(1)} = V_{k-1} + W_{k-1},$$

we get

$$A + P_1 + \ldots + P_k = f(Q_k) + A^{(1)} + P_1^{(1)} + \ldots + P_{k-1}^{(1)},$$

and hence, by (36),

$$f(B + Q_1 + \ldots + Q_k) = A^{(1)} + P_1^{(1)} + \ldots + P_{k-1}^{(1)} + f(Q_k).$$

This implies

$$f(B + Q_1 + \ldots + Q_{k-1}) = A^{(1)} + P_1^{(1)} + \ldots + P_{k-1}^{(1)},$$

i.e., formula (34) holds. Moreover,

$$P_i^{(1)} = V_i + W_i \underset{G}{\sim} V_i + U_i = P_i \underset{G}{\sim} Q_i, \quad i = 1, \ldots, k-1;$$
$$A = (f(Q_k) \cap A) + N + W_1 + \ldots + W_k \underset{G}{\sim}$$
$$\underset{G}{\sim} (f(Q_k) \cap A) + N + U_1 + \ldots + U_k =$$
$$= N + (f(Q_k) \cap A) + (f(Q_k) \cap P_1) + \ldots + (f(Q_k) \cap P_k) =$$

$$= N + (f(Q_k) \cap (A + P_1 + \ldots + P_k)) =$$
$$= N + f(Q_k) \underset{G}{\sim} N + P_k =$$
$$= N + U_k + V_k \underset{G}{\sim} N + W_k + V_k = A^{(1)},$$

so that the formulas (35) also hold.

Proof of Theorem 22. If the polyhedra arc G-equicomplementable, then they are G-equidecomposable (Theorem 11). Conversely, suppose the polyhedra A and B are G-equicomplementable. Then there exist polyhedra $P_1, \ldots, P_k, Q_1, \ldots, Q_k$, such that $P_i \underset{G}{\cong} Q_i$ and

$$A + P_1 + \ldots + P_k \underset{G}{\cong} B + Q_1 + \ldots + Q_k.$$

By dividing the polyhedra P_i, Q_i into smaller pieces if necessary (and thereby increasing their number), we can see to it that the inequalities $2v(P_i) < v(A)$ hold for all $i = 1, \ldots, k$. Then, applying Lemma 13, we find that there exist polyhedra $A^{(1)}$, $P_1^{(1)}$, $\ldots$, $P_{k-1}^{(1)}$ satisfying formulas (34) and (35), where, as before, $2v(P_i^{(1)}) < v(A^{(1)})$, $i = 1, \ldots, k-1$ (since $v(A^{(1)}) = v(A)$, $v(P_i^{(1)}) = v(P_i)$). Applying Lemma 13 once again, we obtain polyhedra $A^{(2)}$, $P_1^{(2)}, \ldots, P_{k-2}^{(2)}$ such that

$$A^{(2)} + P_1^{(2)} + \ldots + P_{k-2}^{(2)} \underset{G}{\cong} B + Q_1 + \ldots + Q_{k-2},$$
$$P_1^{(2)} \underset{G}{\sim} Q_1, \ldots, P_{k-2}^{(2)} \underset{G}{\sim} Q_{k-2}, \quad A^{(2)} \underset{G}{\sim} A^{(1)} \underset{G}{\sim} A.$$

Continuing this process, we get a sequence of relations

$$A^{(j)} + P_1^{(j)} + \ldots + P_{k-j}^{(j)} \underset{G}{\cong} B + Q_1 + \ldots + Q_{k-j}, \quad A^{(j)} \underset{G}{\sim} A^{(j-1)}.$$
$$\tag{39}$$

For $j = k$ the first of the relations (39) takes the form $A^{(k)} \underset{G}{\cong} B$. Thus, taking account of the second relation (39), we find that

$$A \underset{G}{\sim} A^{(1)} \underset{G}{\sim} A^{(2)} \underset{G}{\sim} \ldots \underset{G}{\sim} A^{(k)} \underset{G}{\cong} B,$$

and hence $A \underset{G}{\sim} B$, as required.

The proof of Theorem 22 is set-theoretic in nature,[19] and the only "Euclidean" part of our argument is in the proof of Lemma 12 (on which the theorem is based). But if we replace the relation $G \supset T$ by the requirement that the group G be transitive (cf. Lemma 7, p. 87), then Lemma 12 remains valid (with another proof, of course) in n-dimensional hyperbolic and elliptic spaces, i.e., Lemma 12 and Theorem 22 hold in the following modified forms:

Lemma 12′. *Let G be a transitive group of motions in n-dimensional (Euclidean, hyperbolic or elliptic) space, and let A, B be two polyhedra in this space satisfying the condition $v(A) > v(B)$. Then there exists a decomposition $A = M + N$ such that $M \underset{G}{\sim} B$.*

Theorem 22′. *Let G be a transitive group of motions in n-dimensional (Euclidean, hyperbolic or elliptic) space. Then two polyhedra in this space are G-equicomplementable if and only if they are G-equidecomposable.*

Moreover, the requirement that the group G be transitive can be weakened. In fact, we need only require that G be *almost transitive* (i.e., given any points a, b and any $\varepsilon > 0$, there exists a motion $f \in G$ such that the distance between the points a and $f(b)$ is less than ε). The proof of these results due to Zylev, i.e., the proof of Lemma 12′ for an almost transitive group G (the proofs of Lemma 13 and Theorem 22 remain literally the same) is comparatively simple, but will not be given here.

We now consider some lemmas which constitute the "technique" of equidecomposability, as developed by Sydler [50] and Hadwiger [22], [26], [28]. We will derive some of the lemmas with the help of Theorem 22 (although they were originally proved directly and

---

[19] We note that the proof uses the inequalities $2v(P_i) < v(A)$ for volumes, which hold only insofar as volume is a *real* number, the Archimedean axiom being valid in the field of real numbers. As the results of §8 show, the Archimedean axiom is *essential* for the validity of Theorem 22.

were used by Sydler and Hadwiger as a tool for establishing this theorem).

Given k convex sets $A_1, \ldots, A_k$ in R^n, let $A_1 \times \ldots \times A_k$ denote the *Minkowski sum* of the sets, namely the set of all points of the form $a_1 + \ldots + a_k$, where $a_1 \in A_1, \ldots, a_k \in A_k$ (we are thinking of *vector* sums $a_1 + \ldots + a_k$; i.e., $a_1, \ldots, a_k$ are regarded as elements of the space R^n thought of as a *vector space*). We note that if the sets $A_1, \ldots, A_k$ are subjected to translations, then the Minkowski sum $A_1 \times \ldots \times A_k$ also undergoes a translation. A polyhedron $A \subset R^n$ is called a *k-prism* if there exists an expansion $R^n = L_1 + \ldots + L_k$ of the space R^n as a direct sum of subspaces, together with polyhedra $A_1 \subset L_1, \ldots, A_k \subset L_k$, each of *positive* dimension, such that $A = A_1 \times \ldots \times A_k$. "Ordinary" prisms are 2-prisms in R^3 (Fig. 70). By an *n-parallelepiped* we mean a polyhedron which is an *n*-prism in R^n. Note that if $k > l$, then every *k*-prism is also an *l*-prism. By Z_k we mean the set of all *n*-dimensional polyhedra $M \subset R^n$ which are T-equidecomposable with polyhedra of the form $M_1 + \ldots + M_s$, where s is a positive integer and the polyhedra $M_1, \ldots, M_s$ are *k*-prisms.

Now let $e_1, \ldots, e_k$ be linearly independent vectors of the space R^n, and let $[e_1, \ldots, e_k]$ denote the *k*-dimensional simplex with vertices

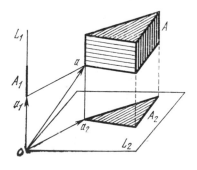

FIG. 70

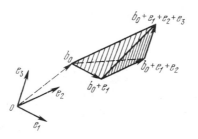

FIG. 71

$$b_0, \quad b_1 = b_0 + e_1, \quad b_2 = b_0 + e_1 + e_2, \ldots$$
$$\ldots, b_k = b_0 + e_1 + \ldots + e_k \quad (40)$$

(Fig. 71). Here b_0 is an arbitrary point of the space R^n, which we call the *initial vertex* of the simplex $[e_1, \ldots, e_k]$. The simplex $[e_1, \ldots, e_k]$ undergoes a translation if its initial vertex is changed. Every point x of the simplex $[e_1, \ldots, e_k]$ with initial vertex b_0 (see (40)) has a unique representation of the form

$$x = (1 - \lambda_1 - \ldots - \lambda_k)b_0 + \lambda_1 b_1 + \ldots + \lambda_k b_k,$$

where $\lambda_1, \ldots, \lambda_k$ are nonnegative numbers satisfying the condition $\lambda_1 + \ldots + \lambda_k \leqslant 1$. In other words,

$$x = b_0 + x_1 e_1 + \ldots + x_k e_k, \quad (41)$$

where

$$x_1 = \lambda_1 + \ldots + \lambda_k, \quad x_2 = \lambda_2 + \ldots + \lambda_k, \quad \ldots$$
$$\ldots, x_{k-1} = \lambda_{k-1} + \lambda_k, \quad x_k = \lambda_k.$$

Thus every point $x \in [e_1, \ldots, e_k]$ has a unique representation of the form (41), where the coordinates $x_1, \ldots, x_k$ are connected by the inequalities

$$1 \geqslant x_1 \geqslant x_2 \geqslant \ldots \geqslant x_k \geqslant 0.$$

Next let $M_1, \ldots, M_p$ be polyhedra, with pairwise disjoint interiors, each of which is T-congruent to a given polyhedron M. In this case, we agree to denote the polyhedron $M_1 + \ldots + M_p$ by $p \cdot M$. Moreover, let λM (without the dot!) denote the polyhedron homothetic to the polyhedron M, with homothetic ratio λ. It is obvious that $\lambda \, [e_1, \ldots, e_k] = [\lambda e_1, \ldots, \lambda e_k]$.

Lemma 14. *The expansion*

$$(\lambda + \mu) [e_1, \ldots, e_n] \underset{T}{\sim} \lambda [e_1, \ldots, e_n] +$$
$$+ \mu [e_1, \ldots, e_n] + \sum_{i=1}^{n-1} \lambda [e_1, \ldots, e_i] \times \mu [e_{i+1}, \ldots, e_n]$$

holds for arbitrary positive λ and μ.

PROOF. The simplex $(\lambda + \mu) [e_1, \ldots, e_n]$ consists of the points

$$x = b_0 + x_1 e_1 + \ldots + x_n e_n \qquad (42)$$

satisfying the condition $\lambda + \mu \geqslant x_1 \geqslant \ldots \geqslant x_n \geqslant 0$. Let M_i $(0 \leqslant i \leqslant n)$ denote the polyhedron consisting of the points (42) for which

$$\lambda + \mu \geqslant x_1 \geqslant \ldots \geqslant x_i \geqslant \mu \geqslant x_{i+1} \geqslant \ldots \geqslant x_n \geqslant 0. \qquad (43)$$

The polyhedra $M_0, M_1, \ldots, M_n$ have pairwise disjoint interiors, and their union coincides with $(\lambda + \mu) [e_1, \ldots, e_n]$, i.e.,

$$(\lambda + \mu) [e_1, \ldots, e_n] = M_0 + M_1 + \ldots + M_n.$$

The polyhedron M_0 consists of the points (42) for which

$$\mu \geqslant x_1 \geqslant \ldots \geqslant x_n \geqslant 0$$

and is just the simplex $[\mu e_1, \ldots, \mu e_n] = \mu [e_1, \ldots, e_n]$ with initial vertex b_0. The polyhedron M_n consists of the points (42) for which $\lambda + \mu \geqslant x_1 \geqslant \ldots \geqslant x_n \geqslant \mu$ and is just the simplex $[\lambda e_1, \ldots, \lambda e_n] = \lambda [e_1, \ldots, e_n]$ with initial vertex $b_0 + \mu e_1 + \ldots + \mu e_n$. Finally, if $0 < i < n$, then every point $x \in M_i$ can be represented in the form $a_1 + a_2$, where

$$a_1 = b_0 + x_1 e_1 + \ldots + x_i e_i, \quad a_2 = x_{i+1} e_{i+1} + \ldots + x_n e_n,$$

and the numbers $x_1, \ldots, x_n$ satisfy the condition (43). From this it is clear that $M_i = A_i \times B_i$, where A_i is the simplex $[\lambda e_1, \ldots, \lambda e_i] = \lambda [e_1, \ldots, e_i]$ with initial vertex $b_0 + \mu e_1 + \ldots + \mu e_i$, and B_i is the simplex $[\mu e_{i+1}, \ldots, \mu e_n] = \mu [e_{i+1}, \ldots, e_n]$ with initial vertex o. The required expansion is now proved. (As an illustration, the case $n = 3$ is shown in Fig. 72.)

Lemma 15. *Let $M \subset R^n$ be any n-dimensional polyhedron, and let λ, μ be arbitrary positive numbers. Then*

$$(\lambda + \mu) M \underset{T}{\sim} \lambda M + \mu M + P,$$

where $P \in Z_2$.

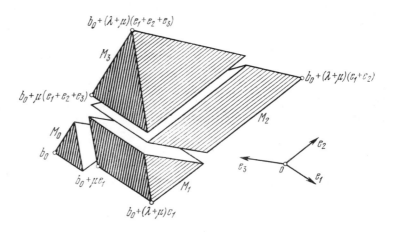

FIG. 72

PROOF. Let $M = B_1 + \ldots + B_k$, where $B_1, \ldots, B_k$ are simplexes. By Lemma 14,

$$(\lambda \dashv \mu) B_j \underset{T}{\sim} \lambda B_j + \mu B_j + P_j, \quad P_j \in Z_2; \quad j = 1, \ldots k.$$

Therefore

$$(\lambda + \mu) M \underset{T}{\sim} \lambda B_1 + \ldots + \lambda B_k + \mu B_1 + \ldots + \mu B_k +$$
$$+ P_1 + \ldots + P_k \underset{T}{\sim} \lambda M + \mu M + P,$$

where $P = P_1 + \ldots + P_k \in Z_2$.

Lemma 16. *Let $M \subset R^n$ be any n-dimensional polyhedron, and let p be an arbitrary positive integer. Then*

$$pM \underset{T}{\sim} p \cdot M + P,$$

where $P \in Z_2$.

PROOF. By Lemma 15, we have $pM \underset{T}{\sim} M + (p - 1) M + P$, where $P \in Z_2$. The rest of the proof follows by an obvious induction.

Now let $\Pi \subset R^n$ be the infinite prism given by the Minkowski sum $M \times l$, where M is an $(n - 1)$-dimensional polyhedron and l is a straight line which does not lie in the hyperplane of M. Moreover, let W be the slab in R^n bounded by two parallel hyperplanes Γ_1 and Γ_2, each of which intersects l in a single point. Then $\Pi \cap W$ is a prism in R^n (in general, oblique) of *lateral edge* length $l \cap W$ with the polyhedra $\Pi \cap \Gamma_1$ and $\Pi \cap \Gamma_2$ as its *bases*.

Lemma 17. *Let $\Pi = M \times l$ be an infinite prism, and let $A_1 = \Pi \cap W_1$, $A_2 = \Pi \cap W_2$ be two prisms cut from Π by slabs W_1, W_2. Suppose the lateral edges of A_1 and A_2 are of the same length. Then A_1 and A_2 are T-equidecomposable.*

PROOF. We can assume (by making a translation in the direction of l, if necessary) that $A_1 \cap A_2 = \varnothing$. Let B denote the part of the infinite prism Π between A_1 and A_2. The polyhedra $A_1 + B$

and $A_2 + B$ are T-congruent (cf. Fig. 21). Therefore, A_1 and A_2 are T-equicomplementable, since

$$A_1 + (A_2 + B) = A_2 + (A_1 + B).$$

But then the polyhedra A_1 and A_2 are T-equidecomposable, by Theorem 22.

Lemma 18. *Any two polyhedra* M, $N \in Z_n$ *in* R^n *with the same (n-dimensional) volume are T-equidecomposable.*

PROOF. We fix a system of rectangular coordinates $x_1, \ldots, x_n$ in R^n. Let A be an arbitrary parallelepiped in R^n, and let I be any edge of A which is not parallel to the hyperplane $x_1 = 0$. According to Lemma 17, the parallelepiped A is T-equidecomposable with the parallelepiped A_1 whose lateral edge lies in the line of I and has the same length as I, but whose bases are parallel to the hyperplane $x_1 = 0$ (Fig. 73). The hyperplanes containing all the $(n - 1)$-dimensional faces other than the bases do not change in going from A to A_1.

We can now replace A_1 by a parallelepiped A_2, which is T-equidecomposable with A_1 (and with A) and has a face parallel to the hyperplane $x_2 = 0$, as well as a face parallel to the hyperplane $x_1 = 0$. Continuing in this way, we eventually obtain a parallelepiped A_n, whose faces are all parallel to the coordinate hyperplanes,

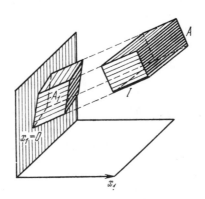

FIG. 73

i.e., whose edges are all parallel to the coordinate axes.

Let $l_1 , \ldots, l_n$ be the lengths of the edges of the parallelepiped A_n, which are parallel to the axes $x_1 , \ldots, x_n$. By Lemma 3_T, p. 76, we can replace A_n by a parallelepiped P such that $P \underset{T}{\sim} A_n$, whose edges parallel to the $x_3 , \ldots, x_n$ axes have the same lengths $l_3 , \ldots, l_n$ as before, but whose edges parallel to the x_1 and x_2 axes now have lengths 1 and $l_1 l_2$ (Fig. 74). Continuing in this way, we can make all the lengths of the edges parallel to the $x_1, x_2, \ldots, x_{n-1}$ axes equal to 1. Thus, finally, we see that every parallelepiped is T-equidecomposable with a *rectangular* parallelepiped, whose base is a unit cube in the hyperplane $x_n = 0$.

Now let $M \in Z_n$, so that $M \underset{T}{\sim} M_1 + \ldots + M_k$, where $M_1, \ldots, M_h$ are parallelepipeds. Replacing $M_1, \ldots, M_k$ by rectangular parallelepipeds $P_1 , \ldots, P_k$, where $P_1 \underset{T}{\sim} M_1 , \ldots, P_k \underset{T}{\sim} M_k$, whose bases are unit cubes in the hyperplane $x_n = 0$, and then "piling up" $P_1, \ldots, P_k$, as shown in Fig. 75, we get a parallelepiped $P = P_1 + \ldots + P_k$, which is T-equidecomposable with M. The base of the parallelepiped P is a unit cube in the hyperplane $x_n = 0$, and its altitude is equal to $v(M)$. Thus two polyhedra M, $N \in Z_n$ with the same volume are T-equidecomposable with one and the same parallelepiped P, and therefore $M \underset{T}{\sim} N$ as asserted.

Lemma 19. *Any two polyhedra M, $N \in Z_2$ in R^3 with the same volume are equidecomposable.*

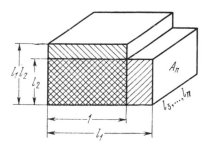

FIG. 74

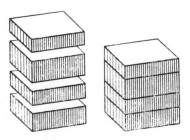

FIG. 75

PROOF. Let $M \in Z_2$, so that $M \underset{T}{\sim} M_1 + \ldots + M_k$, where $M_1, \ldots, M_k$ are prisms. By Lemma 17, every prism M_i is equidecomposable with some *right* prism. This right prism can be decomposed into a number of right triangular prisms (Fig. 24), and each of these right triangular prisms is equidecomposable with a rectangular parallelepiped, as shown in Fig. 76 (cf. Fig. 28). Thus $M \sim P$, where $P \in Z_3$, and the rest of the proof follows from Lemma 18.

Lemmas 16 and 19 were used by Sydler [50] to prove that *equicomplementable three-dimensional polyhedra are equidecomposable.* For the sake of comparison, we now give his proof. Let A and B be equicomplementable, i.e.,

$$A + C \cong B + D, \quad C \sim D, \tag{44}$$

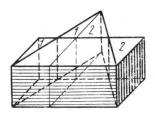

FIG. 76

and let p be a positive integer satisfying the condition $p^2v\,(A) >$ $v\,(A) + v\,(C)$. Then

$$v\,(pA) = p^3v\,(A) > pv\,(A) + pv\,(C) = v\,(p \cdot A + p \cdot C). \quad (45)$$

Moreover, according to Lemma 16,

$$pA \sim p \cdot A + P, \quad pB \sim p \cdot B + Q, \quad (46)$$

where $P, Q \in Z_2$. It follows from (45) and (46) that $v\,(P) >$ $v\,(p \cdot C)$, and hence, by Lemma 12, there exist polyhedra M and N satisfying the conditions

$$P = M + N, \quad M \sim p \cdot C. \quad (47)$$

Moreover, (46) implies $v\,(P) = v\,(Q)$, since $v\,(pA) = v\,(pB)$, $v\,(p \cdot A) = v\,(p \cdot B)$. Therefore, according to Lemma 19,

$$P \sim Q. \quad (48)$$

From (44), (46)–(48) we obtain

$$pA \sim p \cdot A + P = p \cdot A + M + N \sim p \cdot A + p \cdot C + N =$$
$$= p \cdot (A + C) + N \sim p \cdot (B + D) + N = p \cdot B +$$
$$+ p \cdot D + N \sim p \cdot B + p \cdot C + N \sim p \cdot B + M +$$
$$+ N = p \cdot B + P \sim p \cdot B + Q \sim pB.$$

Thus pA and pB are equidecomposable, i.e., they have decompositions in which corresponding pieces are congruent. Subjecting the polyhedra A and B, as well as the pieces making them up, to a similarity transformation with ratio $1/p$, we finally get $A \sim B$.

In conclusion, we consider a problem connected with the work of Zylev [60], which is close to the theme of this section. Let Γ denote the group consisting of all translations of the space R^n and all homothetic transformations with positive ratios, and let G be a group of affine transformations of R^n containing Γ. Then an arbitrary transformation $g \in G$ carries polyhedra into polyhedra, and hence we can talk about G-equidecomposability and G-equicomplementability of polyhedra in R^n. The following theorems are due to Zylev [60].

Theorem 23. *Let G be a group of affine transformations of the space R^n containing Γ. Then two polyhedra A and B in R^n are G-equicomplementable if and only if they are G-equidecomposable.*

PROOF. First we prove the following analogue of Lemma 12: Given any two polyhedra A, $B \subset R^n$, there exists a decomposition $A = M + N$ such that $M \underset{G}{\cong} B$. In fact, let S_0 and S_1 be balls such that $S_0 \supset B$, $S_1 \subset A$. Then there exists a homothetic transformation (or a translation) $g \in \Gamma \subset G$ carrying S_0 into S_1. Thus $g(B) \subset A$, and we need only set $M = g(B)$, $N = \overline{A \setminus M}$.

The arguments used to prove Lemma 12 and Theorem 22 now carry over without changes (except that the inequalities $2v(P_i) < v(A)$ are no longer necessary).

Theorem 24. *Let Π denote the group of all similarity transformations in R^3 (i.e., compositions of homothetic transformations and motions). Then any two polyhedra in R^3 are Π-equidecomposable.*

PROOF. Let M be any polyhedron in R^3. By Lemma 16,

$$M \underset{\Pi}{\cong} 3M \underset{T}{\sim} 3 \cdot M + P_1, \quad M \underset{\Pi}{\cong} 2M \underset{T}{\sim} 2 \cdot M + Q,$$

where $P_1, Q \in Z_2$. Thus

$$3 \cdot M + P_1 \underset{\Pi}{\sim} 2 \cdot M + Q,$$

i.e., there exist polyhedra P_2, P_3, Q_2, Q_3, congruent to M, such that

$$(M + P_2 + P_3) + P_1 \underset{\Pi}{\sim} (Q_2 + Q_3) + Q.$$

Now let A be a cube, and let $\lambda > 0$ be the number such that $v(\lambda Q) = v(A) + v(P_1)$. Then

$$Q \underset{\Pi}{\cong} \lambda Q \underset{D}{\sim} A + P_1$$

(by Lemma 19), and therefore

$$M + P_1 + P_2 + P_3 \underset{\Pi}{\sim} A + P_1 + Q_2 + Q_3,$$

where $P_2 \cong Q_2$, $P_3 \cong Q_3$, i.e., the polyhedra M and A are Π-equicomplementable and hence Π-equidecomposable (by Theorem 23). Thus we see that every polyhedron in R^3 is Π-equidecomposable with a cube. Therefore, any two polyhedra in R^3 are Π-equidecomposable.

We note that the following much stronger result holds in the plane: *Any two polygons are Γ-equidecomposable.* The proof is the same, except that P_i, Q_i and Q are now *parallelograms*, and instead of Lemma 19, we resort to the fact that any two parallelograms are Γ-equidecomposable (by subjecting one of them to a homothetic transformation, we can make their areas equal, and then they are T-equidecomposable by Theorem 16, or by Lemma 18). Figure 77 illustrates the Γ-equidecomposability of a triangle and a rectangle; Γ-congruent polygons are marked with the same number, and ξ

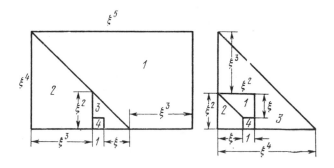

FIG. 77

denotes the positive root of the equation $\xi^2 = \xi + 1$, i.e., $\xi = \frac{1}{2}(\sqrt{3}+1)$.

It is not known whether or not two arbitrary polyhedra in R^3 are Γ-congruent. It is also not known whether or not Theorem 24 is valid in R^n for $n > 3$.

§17. The Dehn–Sydler theorem

The theory of equidecomposability was enriched by a remarkable new result in 1965, when Sydler [54] proved that Dehn's necessary condition (see Theorem 19) is also sufficient, i.e., *a necessary and sufficient condition for equidecomposability of two polyhedra A and B of equal volume is that their Dehn invariants $f(A)$ and $f(B)$ be equal for every additive function f satisfying the condition $f(\pi) = 0$.* This result will be called the *Dehn–Sydler* theorem.[20] In 1968 Jessen [36] (see also [38]), while preserving Sydler's basic geometric lemmas, managed to greatly simplify the algebraic part of the proof of the Dehn–Sydler theorem, and Jessen's proof is the one that will be given here.

Let Z_* denote the set of all polyhedra which are equidecomposable with a cube. By Lemma 19, every polyhedron $M \in Z_2$ is equidecomposable with the cube of equal volume, i.e., $Z_2 \subset Z_*$.

Lemma 20. *Let M be a pyramid with base abcd and vertex o, and suppose the edges bc and ad are perpendicular to the plane ocd, with $| cd | = | od |$ and $| bc | = 2 | ad |$. Then $M \in Z_*$.*

PROOF. Let k and l be the midpoints of the edges ob and bc (Fig. 78), and let h be the point symmetric to l with respect to k. Then $\overrightarrow{oh} = \overrightarrow{lb} = \overrightarrow{cl} = \overrightarrow{da}$, i.e., the triangle hla can be obtained

---

[20] In Nicoletti's paper [44] there are necessary conditions for equidecomposability of polyhedra, comprising the Dehn conditions and, as the author supposed, a number of further conditions that do not reduce to Dehn's conditions. However, since the Dehn conditions are also *sufficient* (according to the Dehn–Sydler theorem), Nicoletti's conditions are *equivalent* to Dehn's.

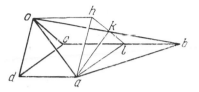

FIG. 78

from the triangle *ocd* by making a translation. Therefore, $| \, ah \, | = | \, al \, |$, and hence $ak \perp hl$. Moreover, the segment *ak* is parallel to the plane *ocd*, and hence $ak \perp bc$, i.e., *ak* is perpendicular to the plane *obc*. Thus a rotation through the angle π about the line *ak* carries the tetrahedron *akbl* into the tetrahedron *akoh*. This means that the pyramid M is equidecomposable with the prism *ocdhla*. Therefore, $M \in Z_*$, as asserted.

To formulate the next lemma, we consider the function

$$w(x) = \frac{1 - x}{x}. \tag{49}$$

For x, $y \in \,]0, \, 1[$,[21] let $T \, (x, \, y)$ denote the tetrahedron *abcd*, where the edges *ab, bc, cd* are mutually perpendicular and have lengths $\sqrt{w \, (x)}$, $\sqrt{w \, (x) \, w \, (y)}$, $\sqrt{w \, (y)}$, respectively. In this tetrahedron the dihedral angle at the edge *ab* equals the angle *cbd* (Fig. 79), i.e., equals $\arctan 1/\sqrt{w \, (x)}$, while the dihedral angle at the edge *cd* equals $1/\sqrt{w \, (y)}$. Since the vectors $(1, \, 0, \, \sqrt{w \, (y)})$ and $(-1, \sqrt{w \, (x)}, \, 0)$ are orthogonal to the planes *acd* and *abd*, respectively (Fig. 80), the dihedral angle at the edge *ad* equals

$$\arccos \frac{1}{\sqrt{1 + w \, (x)} \, \sqrt{1 + w \, (y)}} = \arccos \sqrt{xy} =$$

$$= \frac{\pi}{2} - \arctan \frac{1}{\sqrt{w \, (xy)}}$$

[21] By $]a, \, b[$ is meant the *open* interval with end points a and b $(a < b)$.

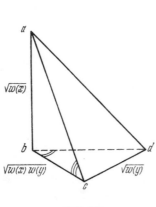

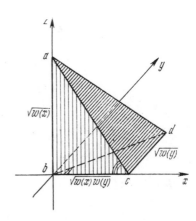

FIG. 79   FIG. 80

(see (49)). Moreover,

$$|ad| = \sqrt{|ab|^2 + |bc|^2 + |cd|^2} =$$
$$= \sqrt{w(x) + w(x)w(y) + w(y)} = \sqrt{w(xy)}, \qquad (50)$$

$$v(T(x, y)) = \frac{1}{6}|ab| \cdot |bc| \cdot |cd| = \frac{1}{6}w(x)w(y) =$$
$$= \frac{1}{6}(w(xy) - w(x) - w(y)). \qquad (51)$$

Lemma 21. *The formula*

$$T(x, y_1) + T(xy_1, y_2) \sim T(x, y_2) + T(xy_2, y_1). \qquad (52)$$

holds for arbitrary x, y_1, $y_2 \in \,]0, 1[$.

PROOF. Figure 81 shows the tetrahedra $abc_1d_1 = T(x, y_1)$ and $abc_2d_2 = T(x, y_2)$, with common edge ab; here the angles c_1bd_1 and c_2bd_2 are equal (their common value being arctan $1/\sqrt{w(x)}$). Since the angles $d_1c_1d_2$ and $d_1c_2d_2$ are right angles, the points c_1, d_1, c_2, d_2 lie on some circle with center at the midpoint g of the segment d_1d_2. Hence the points a, c_1, d_1, c_2, d_2 lie on some sphere

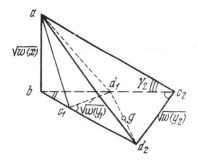

FIG. 81

S, whose center h lies on the perpendicular to the plane bd_1d_2, drawn through the point g. Let M, P, Q, R denote the polyhedra $abd_1d_2hp_1p_2$, $ahp_2c_1d_1$, $d_2hp_2c_1d_1$, ahd_1p_1 (Fig. 82), where p_1 and p_2 are the projections of h onto the planes abd_1 and abd_2. Then

$$M = T\,(x,\,y_1) + P + Q + R. \qquad (53)$$

(We note that with another choice of the numbers x, y_1, y_2 it is possible for the terms P, Q to be transferred to the left-hand side of formula (53); for example, if the point g lies inside the triangle bc_1d_1

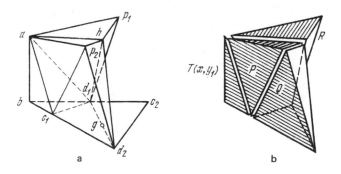

FIG. 82

and if the segments c_1d_1 and c_2d_2 do not intersect, then

$$M + Q = T(x, y_1) + P + R,$$

but this has no effect on the nature of the subsequent argument.)

The plane abc_2 intersects the sphere S in a circle K, with center p_1. Since the points a, c_2, d_1 lie both on the sphere S and in the plane abc_2, they belong to the circle K:

$$|p_1a| = |p_1d_1| = |p_1c_2|.$$

In the same way,

$$|p_2a| = |p_2c_1| = |p_2d_2|.$$

Moreover, since the segment gh is parallel to the plane abc_2, both g and h are at the same distance from the plane, while d_2 is at twice this distance, i.e., the segments p_1h and c_2d_2 are perpendicular to the plane abc_2 and $|c_2d_2| = 2|p_1h|$. Similarly, the segments p_2h and c_1d_1 are perpendicular to the plane abc_1 and $|c_1d_1| = 2|p_2h|$. Therefore, $P \in Z_*$, $Q \in Z_*$, by Lemma 20.

Finally, we deal with the polyhedron R. Let e_1 be the point of the circle K diametrically opposite to a. The inscribed angles ae_1d_1 and ac_2d_1 are equal (Fig. 83), and hence the angle ae_1d_1 equals $\gamma_2 = \arctan 1/\sqrt{w(y_2)}$ (Fig. 81). Since $|ad_1| = \sqrt{w(xy_1)}$ (see (50)), $|d_1e_1| = |ad_1| \cot \gamma_2 = \sqrt{w(xy_1)} \cdot \sqrt{w(y_2)}$. Let q denote the point symmetric to a with respect to the point h (Fig. 84). Then $|e_1q| = 2|p_1h| = |c_2d_2| = \sqrt{w(y_2)}$. Thus the edges ad_1, d_1e_1, e_1q are mutually perpendicular and have lengths $\sqrt{w(xy_1)}$, $\sqrt{w(xy_1)w(y_2)}, \sqrt{w(y_2)}$; i.e., the tetrahedron ad_1e_1q is $T(xy_1, y_2)$. It is clear that $T(xy_1, y_2) = R + N$, where N is the pyramid $d_1hp_1e_1q$. But $|p_1e_1| = |p_1d_1|$ (since p_1 is the center of the circle K), and hence $N \in Z_*$, by Lemma 20. Taking account of (53), we now have

$$M + U \sim T(x, y_1) + T(xy_1, y_2) + V,$$

$$\tag{54}$$

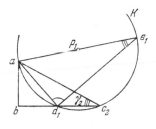

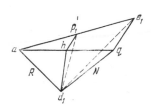

FIG. 83   FIG. 84

where $U, V \in Z_*$ (to obtain this formula, it may be necessary to displace the polyhedra with the help of appropriate motions, so that the terms will have no common interior points). Changing the roles of the subscripts 1 and 2, we find similarly that

$$M + U' \sim T(x, y_2) + T(xy_2, y_1) + V', \qquad (55)$$

where $U', V' \in Z_*$. It follows from (54) and (55) that

$$T(x, y_1) + T(xy_1, y_2) + V + U' \sim M + U + U' \sim$$
$$\sim T(x, y_2) + T(xy_2, y_1) + V' + U. \qquad (56)$$

We now note that the polyhedra appearing in the left- and right-hand sides of formula (52) *have the same volume* (see (51)). Therefore, by (56), $v(V + U') = v(V' + U)$, and hence, the polyhedra $V + U' \in Z_*$ and $V' + U \in Z_*$ are equidecomposable. According to (56), this means that the polyhedra

$$T(x, y_1) + T(xy_1, y_2) \quad \text{and} \quad T(x, y_2) + T(xy_2, y_1)$$

are equicomplementable and hence (by Theorem 22) equidecomposable.

Lemma 22. *The formula*

$$xT\left(\frac{x+y}{x+y+z}, \frac{x}{x+y}\right) + yT\left(\frac{x+y}{x+y+z}, \frac{y}{x+y}\right) \sim$$
$$\sim xT\left(\frac{x+z}{x+y+z}, \frac{x}{x+z}\right) + zT\left(\frac{x+z}{x+y+z}, \frac{z}{x+z}\right)$$

holds for arbitrary $x > 0,\ y > 0,\ z > 0$.

PROOF. Suppose that the edges $oa,\ ob,\ oc$ of the tetrahedron $oabc$ are mutually perpendicular and have lengths $\sqrt{yz},\ \sqrt{xz},\ \sqrt{xy}$. The plane through the edge oc perpendicular to edge ab divides $oabc$ into two tetrahedra $adoc$ and $bdoc$ (Fig. 85). The edges $bd,\ do,\ oc$ are mutually perpendicular and have the lengths

$$x\sqrt{\frac{z}{x+y}},\quad \sqrt{\frac{xyz}{x+y}},\quad \sqrt{xy},$$

from which it follows easily that the tetrahedron $bdoc$ coincides with

$xT\left(\dfrac{x+y}{x+y+z},\ \dfrac{x}{x+y}\right)$. On the other hand, $adoc$ coincides with

$yT\left(\dfrac{x+y}{x+y+z},\ \dfrac{y}{x+y}\right)$. Thus the tetrahedron $oabc$ equals the left-

hand side of the formula to be proved. To get the right-hand side, we need only divide the same tetrahedron $oabc$ into two tetrahedra by drawing the plane through the edge ob perpendicular to the edge ac.

Lemma 23. *Suppose the numbers* $\xi,\ \eta,\ \zeta \in \,]0,\pi/2[$ *satisfy the condition* $\xi + \eta + \zeta = \pi$. *Then there exists a rectangular parallelepiped* R *with diagonals* $ab,\ cd,\ ef,\ gh$ *such that the dihedral angles at the edge* ab *of the tetrahedra* $T^{(1)},\ldots,\ T^{(6)}$ *into which* R *is divided by the planes* $abcd,\ abef,\ abgh$ *are equal to* $\xi,\ \eta,\ \zeta,\ \xi,\ \eta,\ \zeta$.

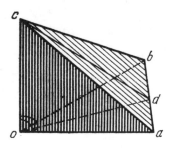

FIG. 85

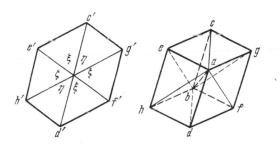

FIG. 86

PROOF. Consider the hexagon $c'g'f'd'h'e'$ which is the affine image of a regular hexagon and has the angles ξ, η, ζ between its diagonals (Fig. 86). Then the rectangular parallelepiped R whose orthogonal projection along the diagonal ab coincides with this hexagon is just the one that we want.

We now turn to Jessen's algebraic considerations [36], [38].

Lemma 24. *Let $F(x, y)$ be a real function, defined for x, $y \in]0, 1[$ which satisfies the conditions*

$$F(x, y) = F(y, x),$$
$$F(x, y_1) + F(xy_1, y_2) = F(x, y_2) + F(xy_2, y_1).$$

Then there exists a real function $f(x)$, defined on the interval $]0, 1[$, such that $F(x, y) = f(x) + f(y) - f(xy)$.

PROOF. We first use the formula $\Phi(u, v) = F(e^{-u}, e^{-v})$ to define a function $\Phi(u, v)$ for all *positive* u and v; this function satisfies the conditions

$$\Phi(u, v) = \Phi(v, u), \quad \Phi(u, v_1) + \Phi(u + v_1, v_2) =$$
$$= \Phi(u, v_2) + \Phi(u + v_2, v_1). \quad (57)$$

We then set

$$\Phi(0, 0) = \Phi(0, v) = \Phi(u, 0) = 0,$$

$$\Phi\left(-u,\, -v\right),\, =\, -\, \Phi\left(u,\, v\right), \qquad (58)$$

$$\Phi\left(u,\, v\right)\, =\, \Phi\left(u,\, w\right)\, =\, \Phi\left(v,\, w\right)\ \text{if}\ \ u + v + w = 0 \qquad (59)$$

and observe that these formulas define the function $\Phi\left(u,\, v\right)$ for all real u and v (since among three nonzero numbers u, v, w satisfying the equation $u + v + w = 0$, there are always two numbers with the same sign).

Next we show that the function $\Phi\left(u,\, v\right)$ just defined satisfies the conditions (57) for *arbitrary* real u, v, v_1, v_2. The validity of the first condition is obvious. To prove the second condition, let

$$G\left(u,\, v_1,\, v_2\right) =$$
$$=\, \Phi\left(u,\, v_1\right)\, +\, \Phi\left(u + v_1,\, v_2\right)\, -\, \Phi\left(u,\, v_2\right)\, -\, \Phi\left(u + v_2,\, v_1\right).$$

Then

$$G\left(u,\, v_1,\, v_2\right)\, =\, -G\left(u,\, v_2,\, v_1\right), \qquad (60)$$

$$G\left(u,\, v_1,\, v_2\right)\, =\, 0\ \ \text{if}\ \ u > 0,\quad v_1 > 0,\, v_2 > 0 \qquad (61)$$

(see (57)). Moreover, according to (58) and (59),

$$G\left(u + v_1,\, -v_1,\, v_2\right)\, =\, \Phi\left(u + v_1,\, -v_1\right)\, +\, \Phi\left(u,\, v_2\right)\, -$$
$$-\, \Phi\left(u + v_1,\, v_2\right)\, -\, \Phi\left(u + v_1 + v_2,\, -v_1\right)\, =$$
$$=\, \Phi\left(-u,\, -v_1\right)\, +\, \Phi\left(u,\, v_2\right)\, -\, \Phi\left(u + v_1,\, v_2\right)\, -$$
$$-\, \Phi\left(-u - v_2,\, -v_1\right)\, =\, -\, G\left(u,\, v_1,\, v_2\right),$$

and hence

$$G\left(u,\, v_1,\, v_2\right)\, =\, -\, G\left(u + v_1,\, -v_1,\, v_2\right). \qquad (62)$$

The formulas

$$G\left(u',\, v',\, v''\right)\, =\, -\, G\left(-u',\, u' + v',\, u' + v''\right)\, =$$
$$=\, G\left(v',\, -u' - v',\, u' + v''\right)\, =$$
$$=\, G\left(v'',\, u' + v',\, -u' - v''\right) \qquad (63)$$

are proved similarly. For example, the first of these formulas can be established as follows

$$G\left(-u', u' + v', u' + v''\right) = \Phi\left(-u', u' + v'\right) +$$
$$+ \Phi(v', u' + v'') - \Phi\left(-u', u' + v''\right) - \Phi\left(v'', u' + v'\right) =$$
$$= \Phi(-u', -v') + \Phi\left(u' + v'', v'\right) - \Phi\left(-u', -v''\right) -$$
$$- \Phi\left(u' + v', v''\right) = -G\left(u', v', v''\right).$$

Moreover, setting $\sigma = u' + v' + v''$, we can prove the formulas

$$G\left(u', v', v''\right) = -G\left(\sigma, -u' - v', -u' - v''\right), \quad (64)$$

$$G\left(u', v', v''\right) = G\left(-\sigma, v', v''\right). \quad (65)$$

Now let u, v_1, v_2 be arbitrary real numbers. Applying formula (62), if necessary, we can see to it that the last two arguments are nonnegative, i.e.,

$$G\left(u, v_1, v_2\right) = \pm G\left(u', v', v''\right); \quad v' \geqslant 0, \quad v'' \geqslant 0.$$

If at least one of the numbers u', v', v'' equals zero, then it can be verified at once that $G\left(u', v', v''\right) = 0$. Therefore, we assume that all the numbers u', v', v'' are nonzero. If $u' = 0$, then $G\left(u', v', v''\right) = 0$, according to (61). Moreover, if $u' < 0$, but at least one of the numbers $u' + v'$, $u' + v''$ is nonnegative, then $G\left(u', v', v''\right) = 0$, because of (63). Finally, if both numbers $u' + v'$, $u' + v''$ are negative, then $G\left(u', v', v''\right) = 0$ also (this follows from (64) if $\sigma \geqslant 0$ and from (65) if $\sigma < 0$). Thus, in any case,

$$G\left(u, v_1, v_2\right) = \pm G\left(u', v', v''\right) = 0,$$

i.e., the validity of the second of the conditions (57) (for arbitrary real arguments) is confirmed.

We now define the operation of addition on the set $R \times R$ of all pairs of real numbers, by setting

$$(a, x) + (b, y) = (a + b, x + y + \Phi\left(a, b\right)). \quad (66)$$

Using (57), we can immediately verify that this operation converts $R \times R$ into an abelian group with zero element $(0, 0)$. Let π denote the operation of projecting the set $R \times R$ onto its first coordinate, i.e., $\pi(a, x) = a$. Then a subgroup $H \subset R \times R$ will be called "marked" if π is *one-to-one* on H, i.e., if $(a, x) \in H$, $(a, x') \in H$ implies $x = x'$. The set M of all marked subgroups is nonempty; for example, the subgroup consisting of the zero element $(0, 0)$ alone is marked. The set M is ordered by inclusion, i.e., $H < H'$ if $H \subset H'$. It is clear that if a set of marked subgroups is a *chain* (see p. 31), then the union of all these subgroups is also a *marked* subgroup. Therefore, by Zorn's lemma, there exists at least one *maximal* marked subgroup.

Let H^* be a maximal marked subgroup. Then, as we now show, $\pi(H^*) = R$. Suppose, to the contrary, that there exists a number $c \notin \pi(H^*)$. Assuming first that the inclusion $kc \in \pi(H^*)$ holds for some integer $k \neq 0$, let q be the smallest positive integer satisfying the condition $qc \in \pi(H^*)$. Then there exists a number b such that $(qc, b) \in H^*$. In fact,

$$q(c, x) = \left(qc, \, qx + \sum_{k=1}^{q-1} \Phi(c, kc)\right)$$

(for every $x \in R$), and therefore, setting

$$d = \frac{1}{q}\left(b - \sum_{k=1}^{q-1} \Phi(c, kc)\right),$$

we find that $q(c, d) = (qc, b) \in H^*$. However, if the inclusion $kc \in \pi(H^*)$ fails to hold for every integer $k \neq 0$, then we set $q = 0$ and choose the number d arbitrarily; in this case we again have $q(c, d) \in H^*$.

Now let H^{**} denote the subgroup consisting of all elements of the form $(a, x) + n(c, d)$, where $(a, x) \in H^*$ and n is an integer. If the element $(a, x) + n(c, d) \in H^{**}$ satisfies the condition $\pi((a, x) + n(c, d)) = 0$, i.e., $a + nc = 0$, then $nc = -a \in \pi(H^*)$, and hence $n = kq$, where k is an integer. Therefore

$$(a, x) + n (c, d) = (a, x) + k \cdot q (c, d) \in H^*. \quad (67)$$

Since the group H^* is marked, it follows from (67) and the formula $\pi ((a, x) + n (c, d)) = 0$ that $(a, x) + n (c, d) = (0, 0)$. Thus (0, 0) is the only element of the subgroup H^{**} that is projected into the point 0, i.e., H^{**} is also a marked subgroup, contrary to the assumed maximality of the subgroup H^* (since $H^{**} \supset H^*$, while $(c, d) \in H^{**}$, $(c, d) \notin H^*$). This contradiction shows that $\pi (H^*) = R$.

The fact that $\pi (H^*) = R$ means that given any $u \in R$, there is a (unique) number $\varphi (u)$ such that $(u, \varphi (u)) \in H^*$. Moreover, given any $u, v \in R$ we have $(u, \varphi (u)) + (v, \varphi (v)) \in H^*$ i.e.,

$$(u + v, \varphi (u) + \varphi (v) + \Phi (u, v)) \in H^*,$$

which implies

$$\varphi (u) + \varphi (v) + \Phi (u, v) = \varphi (u + v).$$

Therefore, setting $f (e^{-u}) = -\varphi (u)$, we finally find that

$$\begin{aligned} F (x, y) = F (e^{-u}, e^{-v}) = \Phi (u, v) = \\ = \varphi (u + v) - \varphi (u) - \varphi (v) = \\ = f (e^{-u}) + f (e^{-v}) - f (e^{-u-v}) = f (x) + f (y) - f (xy) \end{aligned}$$

for every pair of numbers $x = e^{-u}$, $y = e^{-v}$ in the interval $]0, 1[$.

Lemma 25. *Let $\Phi (u, v)$ be a real function, which is defined for $u > 0, v > 0$ and satisfies the conditions*

$$\Phi (u, v) = \Phi (v, u), \quad \Phi (\lambda u, \lambda v) = \lambda \Phi (u, v), \quad (68)$$
$$\Phi (u, v_1) + \Phi (u + v_1, v_2) = \Phi (u, v_2) + \Phi (u + v_2, v_1). \quad (69)$$

Then there exists a real function $g(u)$, which is defined for $u > 0$ and satisfies the condition

$$g (u) + g (v) - g (uv) = 0, \quad (70)$$

such that

$$\Phi\,(u,\,v) = ug\,(u) + vg\,(v) - (u + v)\,g\,(u + v). \quad (71)$$

PROOF. We extend the function $\Phi\,(u,v)$ for all real $u,\,v$ in the same way as in the proof of Lemma 24 (see (58), (59)), so that $\Phi\,(u,v)$ now satisfies the conditions (68) and (69) for all real $u,\,v,\,v_1,\,v_2',\,\lambda$. We then define addition and multiplication in $R \times R$, by setting

$$(a,\,x) + (b,\,y) = (a +_{|} b,\, x + y + \Phi\,(a,\,b)),$$

$$(a,\,x)\,(b,\,y) = (ab,\, bx + ay).$$

These operations convert $R \times R$ into a commutative ring with zero element $(0, 0)$ and unit element $(1, 0)$.

Next let π denote projection onto the first coordinate, i.e., $\pi\,(a,\,x) = a$. Then a subring $H \subset R \times R$, containing the unit element $(1, 0)$, will be called "marked" if π is *one-to-one* on H, i.e., if $(a,\,x) \in H$, $(a,\,x') \in H$ implies $x = x'$. The set M of all marked subrings is nonempty; for example, the set H_0 consisting of all elements of the form $n \cdot (1, 0)$, where n takes all integral values, is a marked subring (since $\pi\,(n \cdot (1, 0)) = n$). The set M is ordered by inclusion, i.e., $H < H'$ if $H \subset H'$. It is clear that if a set of marked subrings is a *chain*, then the union of these subrings is also a *marked* subring. Therefore, by Zorn's lemma, there exists at least one *maximal* marked subring.

Let H^* be a maximal marked subring. Then, as we now show, $\pi\,(H^*) = R$. Suppose, to the contrary, that there exists a number $c \notin \pi\,(H^*)$. The isomorphism π of the ring H^* onto $\pi\,(H^*)$ can be extended to an isomorphism σ of the ring of polynomials $H^*\,[z]$ onto the ring of polynomials $\pi\,(H^*)\,[z]$. First, we consider the case where c is an *algebraic* element over the ring $\pi\,(H^*)$, letting $q \in \pi\,(H^*)\,[z]$ denote the polynomial of lowest degree having c as a root, so that $q\,(c) = 0$. Since $(c,\,x) = (c,\,0) + (0,\,x)$, it follows from Taylor's formula that the polynomial $q_1 = \sigma^{-1}\,(q) \in H^*[z]$ satisfies the condition

$$q_1 ((c, x)) = q_1 ((c, 0)) + q_1' ((c, 0)) (0, x) \qquad (72)$$

(there are no further terms, since $(0, x)^2 = (0, 0)$). But since $\pi (q_1 ((c, 0))) = q (c) = 0$, we have $q_1 ((c, 0)) = (0, b)$, where b is some number, and moreover $q' (c) \neq 0$, since the polynomial q' is of lower degree than q, and hence $q_1' ((c, 0)) = (s, t)$, where $s = q' (c) \neq 0$ and t is some number. Therefore, by (72), $q_1 ((c, x_0)) = (0, 0)$, where $x_0 = -b/s$.

Now let H^{**} denote the subring of the ring $R \times R$ consisting of all elements of the form $p_1 ((c, x_0))$, where $p_1 \in H^*[z]$. If $\pi (p_1 ((c, x_0))) = 0$, then the polynomial $p = \sigma (p_1)$, belonging to the ring $\pi (H^*)$, satisfies the condition $p (c) = 0$. Therefore $ap = qr$, where $a \neq 0$ is an element of the ring $\pi(H^*)$ and $r \in \pi (H^*)[z]$. Since $a \in \pi (H^*)$, we can find a number h such that $(a, h) \in H^*$, and hence $(a, h) p_1 = q_1 r_1$, where $r_1 = \sigma^{-1} (r)$. Therefore $(a, h) p_1 ((c, x_0)) = (0, 0)$, and hence $p_1 ((c, x_0)) = (0, 0)$ (since $a \neq 0$). Thus $(0, 0)$ is the only element of the subring H^{**} that is projected into the point 0, i.e., H^{**} is also a marked subring, contrary to the assumed maximality of the subring H^* (since $H^{**} \supset H^*$, while $(c, x_0) \in H^{**}$, $(c, x_0) \notin H^*$). This contradiction shows that $\pi (H^*) = R$.

On the other hand, suppose c is a *transcendental* element over the ring $\pi (H^*)$, i.e., $p (c) \neq 0$ for every nonzero polynomial $p \in \pi (H^*) [z]$. In this case, we fix the number x_0 arbitrarily and let H^{**} denote the subring consisting of all elements $p_1 ((c, x_0))$ with $p_1 \in H^* [z]$. If $\pi (p_1((c, x_0))) = 0$, i.e., $p (c) = 0$, where $p = \sigma (p_1)$, then $p = 0$, and hence $p_1 = (0, 0)$. In particular, $p_1 ((c, x_0)) = (0, 0)$. Thus the subring H^{**} is marked, and the same contradiction arises, thereby showing that $\pi (H^*) = R$.

The fact that $\pi (H^*) = R$ means that given any $u \in R$, there is a (unique) number $\varphi (u)$ such that $(u, \varphi (u)) \in H^*$. Moreover, given any $u, v \in R$, we have

$$(u, \varphi (u)) + (v, \varphi (v)) \in H^*, \quad (u, \varphi (u)) (v, \varphi (v)) \in H^*,$$

i.e.,

$$(u + v, \; \varphi(u) + \varphi(v) + \Phi(u, v)) \in H^*,$$
$$(uv, \; u\varphi(v) + v\varphi(u)) \in H^*,$$

which implies

$$\varphi(u) + \varphi(v) + \Phi(u, v) = \varphi(u + v),$$
$$u\varphi(v) + v\varphi(u) = \varphi(uv).$$

To complete the proof, we need only set

$$g(u) = - \frac{\varphi(u)}{u} \quad (\text{for } u > 0).$$

Lemma 26. *Let $F(x, y)$ be a real function, which is defined for $x, y \in \,]0, 1[$ and satisfies the conditions*

$$F(x, y) = F(y, x), \; F(x, y_1) + F(xy_1, y_2) =$$
$$= F(x, y_2) + F(xy_2, y_1), \qquad (73)$$

$$xF\left(\frac{x + y_1}{x + y_1 + y_2}, \; \frac{x}{x + y_1}\right) + y_1 F\left(\frac{x + y_1}{x + y_1 + y_2}, \; \frac{y_1}{x + y_1}\right) =$$

$$= xF\left(\frac{x + y_2}{x + y_1 + y_2}, \; \frac{x}{x + y_2}\right) + y_2 F\left(\frac{x + y_2}{x + y_1 + y_2}, \; \frac{y_2}{x + y_2}\right). \qquad (74)$$

Then there exists a real function $h(x)$, which is defined on the interval $]0, 1[$ and satisfies the condition

$$xh(x) + yh(y) = 0 \quad \text{if } x + y = 1, \qquad (75)$$

such that

$$F(x, y) = h(x) + h(y) - h(xy) \qquad (76)$$

for all $x, y \in \,]0, 1[$.

PROOF. It follows from (73), according to Lemma 24, that

$$F(x, y) = f(x) + f(y) - f(xy), \qquad (77)$$

where $f(x)$ is defined on the interval $]0, 1[$. The condition (74) then takes the form

$$\left(xf\left(\frac{x}{x+y_1} \right) + y_1 f\left(\frac{y_1}{x+y_1} \right) \right) + \left((x+y_1) f\left(\frac{x+y_1}{x+y_1+y_2} \right) + \right.$$

$$\left. + y_2 f\left(\frac{y_2}{x+y_1+y_2} \right) \right) = \left(xf\left(\frac{x}{x+y_2} \right) + y_2 f\left(\frac{y_2}{x+y_2} \right) \right) +$$

$$+ \left((x+y_2) f\left(\frac{x+y_2}{x+y_1+y_2} \right) + y_1 f\left(\frac{y_1}{x+y_1+y_2} \right) \right). \quad (78)$$

We now define a function $\Psi(u, v)$ for $u > 0$, $v > 0$, by setting

$$\Psi(u, v) = uf\left(\frac{u}{u+v} \right) + vf\left(\frac{v}{u+v} \right). \quad (79)$$

It follows at once from (79) that

$$\Psi(u, v) = \Psi(v, u), \quad \Psi(\lambda u, \lambda v) = \lambda \Psi(u, v),$$

and (78) can be rewritten in the form

$$\Psi(u, v_1) + \Psi(u+v_1, v_2) = \Psi(u, v_2) + \Psi(u+v_2, v_1).$$

By Lemma 25, there exists a real function $g(u)$, which is defined for $u > 0$ and satisfies the condition (70), such that

$$\Psi(u, v) = ug(u) + vg(v) - (u+v) g(u+v). \quad (80)$$

Finally, we introduce a function $h(x) = f(x) - g(x)$, defined on the interval $]0, 1[$. Then (76) is an immediate consequence of (77) and (70). Moreover, (70) implies $g(1) = 0$, and hence, according to (79) and (80), if $x + y = 1$,

$$\Psi(x, y) = xf(x) + yf(y) = xg(x) + yg(y),$$

from which formula (75) follows at once.

Before formulating the next lemma, let us agree to say that two polyhedra A and B are Z_*-*equivalent* (written $A \overset{*}{\sim} B$) if there exist polyhedra P, $Q \in Z_*$, such that $A + P \sim B + Q$. If $A + B \overset{*}{\sim} C$, we will also write $A \overset{*}{\sim} C - B$. It is easy to see that Z_*-equivalences can be added, i.e., if $A \overset{*}{\sim} B$, $C \overset{*}{\sim} D$, then $A + C \overset{*}{\sim} B + D$ (where it may be necessary to displace the polyhedra to make A and C, as well as B and D, have disjoint interiors). Moreover, by Lemmas 15 and 19, the formula

$$\lambda A + \mu A \overset{*}{\sim} (\lambda + \mu)\, A$$

holds for every polyhedron A and arbitrary λ, μ (if there are negative coefficients, we need only shift terms from one side of the formula to the other). Finally, a set $B = \{N_\alpha\}$ of polyhedra in the space R^3 will be called a *polyhedral basis* if every polyhedron A has a unique representation of the form

$$A \overset{*}{\sim} \sum_\alpha \lambda_\alpha N_\alpha, \tag{81}$$

where the λ_α are real numbers, of which only finitely many are nonzero.

Lemma 27. *There exists a polyhedral basis* $B = \{N_\alpha\}$ *in* R^3.

We will call a set of polyhedra *independent* if none of them can be expressed in terms of the others (cf. (81)). By Zorn's lemma, there exists a *maximal* independent set $B = \{N_\alpha\}$. As we now show, B is a basis.

Let $A \subset R^3$. By adjoining the polyhedron A to the set B, we obtain a set of polyhedra which is no longer independent (since B is maximal), i.e., there is a formula $\nu A \overset{*}{\sim} \sum_\alpha \lambda_\alpha N_\alpha$ expressing A in terms of the N_α, where $\nu \neq 0$ (since otherwise the elements of the set B would be dependent). It can be assumed (by moving terms to the other side of the formula, if necessary) that $\nu > 0$. Then, multiplying by $1/\nu$ (i.e., making a homothetic transformation with ratio $1/\nu$), we get $A \overset{*}{\sim} \sum_\alpha \lambda'_\alpha N_\alpha$, where $\lambda'_\alpha = \lambda_\alpha/\nu$. Thus every polyhedron $A \subset R^3$ can be expressed in terms of the polyhedra N_α.

To show that this representation is unique, suppose there are two representations

$$A \overset{*}{\sim} \sum_\alpha \lambda_\alpha N_\alpha, \quad A \overset{*}{\sim} \sum_\alpha \mu_\alpha N_\alpha. \tag{82}$$

Subtracting the two representations and combining similar terms, we get

$$\sum_\alpha (\lambda_\alpha - \mu_\alpha) N_\alpha \overset{*}{\sim} 0.$$

If some of the numbers $\lambda_\alpha - \mu_\alpha$ are nonzero, then we would find that some of the polyhedra of the set B can be expressed in terms of the others, contrary to the assumed independence of the set B. Thus the representations (82) coincide, i.e., B is a basis.

To formulate the next lemma, we fix a polyhedral basis $B - \{N_\alpha\}$. Then, given any $x, y \in \,]0, 1[$, we express the simplex $T(x, y)$ in terms of elements of B:

$$T(x, y) \overset{*}{\sim} \sum_\alpha F_\alpha(x, y) N_\alpha. \tag{83}$$

It follows from Lemma 21 (because of the expansion (83)) that the function $F_\alpha(x, y)$ satisfies the condition (73) for every α, and hence, by Lemma 22, that $F_\alpha(x, y)$ satisfies the condition (74). Therefore, by Lemma 26, there exists a real function $h_\alpha(x)$, which is defined on the interval $]0, 1[$ and satisfies the condition

$$x h_\alpha(x) + y h_\alpha(y) = 0 \text{ if } x + y = 1, \tag{84}$$

such that

$$F_\alpha(x, y) = h_\alpha(x) + h_\alpha(y) - h_\alpha(xy) \tag{85}$$

for all $x, y \in \,]0, 1[$. We now introduce the function

$$\varphi_\alpha(x) = \begin{cases} \tan x \cdot h_\alpha(\sin^2 x) & \text{if } x \neq \dfrac{k\pi}{2}, \\[2ex] 0 & \text{if } x = \dfrac{k\pi}{2}. \end{cases} \qquad (86)$$

Lemma 28. *The function* $\varphi_\alpha(x)$ *is additive.*

PROOF. First we observe that

$$\varphi_\alpha(x) + \varphi_\alpha(y) = 0 \text{ if } x + y = k\pi/2. \qquad (87)$$

To see this, note that if the numbers x and y are not multiples of $\pi/2$ and k is odd, then $\sin^2 y = \cos^2 x$, $\tan y = \cot x$, and hence

$$\varphi_\alpha(x) + \varphi_\alpha(y) = \frac{1}{\sin x \cos x}(\sin^2 x \cdot h_\alpha(\sin^2 x) + \\ + \cos^2 x \cdot h_\alpha(\cos^2 x)) = 0$$

(see (84)). Moreover, if x and y are not multiples of $\pi/2$, but k is even, then $\sin^2 y = \sin^2 x$, $\tan y = -\tan x$, and (86) implies the validity of (87). Finally, if x and y are multiples of $\pi/2$, then (87) is obvious.

We now introduce the notation

$$\varphi_\alpha(A) = l_1 \varphi_\alpha(\gamma_1) + \ldots + l_m \varphi_\alpha(\gamma_m), \qquad (88)$$

where $l_1, \ldots, l_m$ are the lengths of the sides of the polyhedron A, and $\gamma_1, \ldots, \gamma_m$ are the corresponding dihedral angles. It is not hard to see that

$$\varphi_\alpha(T(x, y)) = F_\alpha(x, y). \qquad (89)$$

In fact, the sides *ab, cd, ad* of the tetrahedron *abcd* shown in Fig. 79 have lengths $l_1 = \sqrt{w(x)}$, $l_2 = \sqrt{w(y)}$, $l_3 = \sqrt{w(xy)}$ (see (50)), and the corresponding dihedral angles are

$$\gamma_1 = \arctan \frac{1}{\sqrt{w(x)}}, \quad \gamma_2 = \arctan \frac{1}{\sqrt{w(y)}},$$

$$\gamma_3 = \frac{\pi}{2} - \arctan \frac{1}{\sqrt{w\,(xy)}}\,.$$

Therefore

$$\sin^2 \gamma_1 = \frac{1}{1 + \text{arccot}^2 \gamma_1} = \frac{1}{1 + w\,(x)} = x,$$

$$\sin^2 \gamma_2 = y, \quad \sin^2 \left(\frac{\pi}{2} - \gamma_3\right) = xy.$$

Thus, by (86), (87), and (85),

$$l_1 \varphi_\alpha\,(\gamma_1) + l_2 \varphi_\alpha\,(\gamma_2) + l_3 \varphi_\alpha\,(\gamma_3) =$$
$$= l_1 \varphi_\alpha\,(\gamma_1) + l_2 \varphi_\alpha\,(\gamma_2) - l_3 \varphi_\alpha\left(\frac{\pi}{2} - \gamma_3\right) = F_\alpha\,(x,\,y).$$

To complete the proof of (89), we need only note that the other dihedral angles of the tetrahedron *abcd* are right angles, and according to (86), $\varphi_\alpha\,(\pi/2) = 0$. Together (89) and (83) imply

$$T\,(x,\,y) \overset{*}{\sim} \sum_\alpha \varphi_\alpha\,(T\,(x,\,y))\,N_\alpha. \tag{90}$$

Now suppose the numbers $\xi, \eta, \zeta \in \,]0, \pi/2\,[$ satisfy the condition $\xi + \eta + \zeta = \pi$. Each of the tetrahedra $T^{(1)},\ldots, T^{(6)}$ figuring in Lemma 23 is of the form $\lambda T\,(x,\,y)$ (for suitable $\lambda,\,x,\,y$), and hence, by (90),

$$T^{(i)} \overset{*}{\sim} \sum_\alpha \varphi_\alpha\,(T^{(i)})\,N_\alpha, \qquad i = 1,\,\ldots,\,6. \tag{91}$$

But $T^{(1)} + \ldots + T^{(6)}$ is a parallelepiped, i.e., $T^{(1)} + \ldots + T^{(6)} \overset{*}{\sim} 0$, and therefore

$$\varphi_\alpha\,(T^{(1)}) + \ldots + \varphi_\alpha\,(T^{(6)}) = 0 \tag{92}$$

(see (91)). The dihedral angles of the tetrahedra $T^{(i)}$ at all the edges other than *ae, ag, bf, bh, bc, ad, ab* (see Fig. 86) are *right* angles, and hence the corresponding terms $l_i \varphi_\alpha (\gamma_i)$ equal zero (see (86)). Moreover, each of the edges *ae, ag, bf, bh, bc, ad* is adjoined by two tetrahedra $T^{(1)}, \ldots, T^{(6)}$, and the sum of the dihedral angles at each such edge equals $\pi/2$. Therefore, by (87), the sum of the terms $l_i \varphi_\alpha (\gamma_i)$ corresponding to each such edge equals zero, and nothing remains in the left-hand side of (92) except the terms involving the dihedral angles at the edge *ab*. Hence, by Lemma 23, equation (92) takes the form $2l\varphi_\alpha (\xi) + 2l\varphi_\alpha (\eta) + 2l\varphi_\alpha (\zeta) = 0$, where l is the length of the edge *ab*.

   Thus

$$\varphi_\alpha (\xi) + \varphi_\alpha (\eta) + \varphi_\alpha (\zeta) = 0$$

if ξ, η, $\zeta \in]0, \frac{\pi}{2}[$, $\xi + \eta + \zeta = \pi$. Because of the formula $\varphi_\alpha (\pi - \zeta) = - \varphi_\alpha (\zeta)$ (see (86)), this means that

$$\varphi_\alpha (\xi) + \varphi_\alpha (\eta) = \varphi_\alpha (\xi + \eta) \qquad (93)$$

if $\xi, \eta \in]0, \frac{\pi}{2}[$, $\xi + \eta \in] \frac{\pi}{2}, \pi[$. Formula (93) also holds for $\xi, \eta \in [0, \pi/2]$, $\xi + \eta \in [\pi/2, \pi]$, because of (86) and (87). If $\xi, \eta \in [0, \pi/2]$ and $\xi + \eta \in [0, \pi/2]$, then $\xi' = \pi/2 - \xi, \eta' = \pi/2 - \eta$ satisfy the conditions $\xi', \eta' \in [0, \pi/2], \xi' + \eta' \in [\pi/2, \pi]$, and then, according to (87) and (93),

$$\varphi_\alpha (\xi) + \varphi_\alpha (\eta) = -(\varphi_\alpha (\xi') + \varphi_\alpha (\eta')) =$$
$$= -\varphi_\alpha (\xi' + \eta') = -\varphi_\alpha (\pi - \xi - \eta) = \varphi_\alpha (\xi + \eta).$$

Thus (93) holds for arbitrary $\xi, \eta \in [0, \pi/2]$. Finally, for arbitrary $\xi, \eta \in R$ we have (for suitable integers k, m) $\xi^* = k\pi/2 - \xi \in [0, \pi/2]$, $\eta^* = m\pi/2 - \eta \in [0, \pi/2]$, and hence, by (87),

$$\varphi_\alpha (\xi) + \varphi_\alpha (\eta) = - \varphi_\alpha (\xi^*) - \varphi_\alpha (\eta^*) =$$
$$= - \varphi_\alpha (\xi^* + \eta^*) = - \varphi_\alpha \left(\frac{(k + m) \pi}{2} - \xi - \eta \right) = \varphi_\alpha (\xi + \eta),$$

which proves the lemma.

By (86), the additive function $\varphi_\alpha(x)$ satisfies the condition $\varphi_\alpha(\pi) = 0$. Therefore $\varphi_\alpha(A)$ (see (88)) is a *Dehn invariant of the polyhedron A.*

Lemma 29. *The formula*

$$A \overset{*}{\sim} \sum_\alpha \varphi_\alpha(A) N_\alpha \tag{94}$$

holds for every polyhedron $A \subset R^3$.

We have already seen (cf. (90)) that this formula holds for $T(x, y)$, and hence for every tetrahedron $\mu T(x, y)$. Moreover, if $A \overset{*}{\sim} B$, i.e., $A + P \sim B + Q$ where P, $Q \in Z_*$, then $\varphi_\alpha(A) = \varphi_\alpha(B)$, since $\varphi_\alpha(P) = 0, \varphi_\alpha(Q) = 0$. We also note that if $A \overset{*}{\sim} B \pm C$, where B and C satisfy (94), then A also satisfies (94):

$$A \overset{*}{\sim} B \pm C \overset{*}{\sim} \sum_\alpha (\varphi_\alpha(B) \pm \varphi_\alpha(C)) N_\alpha = \sum_\alpha \varphi_\alpha(A) N_\alpha.$$

Now let $A \subset R^3$ be an arbitrary polyhedron. Then it is easy to see that the formula

$$A \overset{*}{\sim} \sum_i \mu_i T(x_i, y_i) \tag{95}$$

holds. In fact, we first decompose A into tetrahedra and then decompose each tetrahedron into 12 more tetrahedra (by using the center of the inscribed sphere and the points at which the sphere is tangent to the faces), each of which has an edge as an altitude. Consider one of these tetrahedra, say oa_1bc, where the edge oa_1 is perpendicular to the face a_1bc (Fig. 87). Dropping a perpendicular a_1h onto the line bc, we find that the tetrahedron oa_1bc is of the form $\lambda' T(x', y') + \lambda'' T(x'', y'')$, since each of the tetrahedra oa_1hb, oa_1hc is of the form $\lambda T(x, y)$; the signs of the numbers λ', λ'' depend on the position of the point h on the line bc (see Fig. 88). This gives formula (95), from which the validity of formula (94) follows, because of (90).

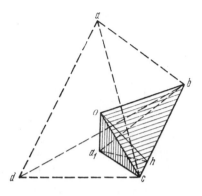

FIG. 87

Theorem 25 (see [12], [54]). *A necessary and sufficient condition for equidecomposability of two polyhedra A and B (in R³) of equal volume is that their Dehn invariants f (A) and f (B) be equal for every additive function f satisfying the condition f (π) = 0.*

PROOF. The necessity is established in Theorem 19. To prove the sufficiency, suppose A and B satisfy the indicated condition. Then, in particular, $\varphi_\alpha (A) = \varphi_\alpha (B)$ for every α. Therefore, by Lemma 29,

$$A \overset{*}{\sim} \sum_\alpha \varphi_\alpha (A) N_\alpha = \sum_\alpha \varphi_\alpha (B) N_\alpha \overset{*}{\sim} B.$$

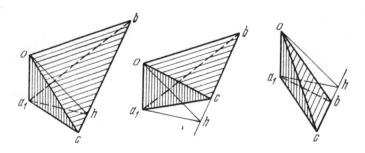

FIG. 88

In other words, $A + P \sim B + Q$, where $P, Q \in Z_*$. Since $v(A) = v(B)$, it follows from $A + P \sim B + Q$ that $v(P) = v(Q)$, so that the polyhedra P and Q are equidecomposable. But then A and B are equicomplementable, and hence equidecomposable, by Theorem 22.

Remark. The proof of the Dehn–Sydler theorem can be modified so that it does not rely on the axiom of choice. To this end, let Y be the set consisting of all the numbers x_i, y_i figuring in formula (95), together with the analogous numbers for the polyhedron B, and let R^* be the field containing the real numbers obtained from the elements of the set Y by carrying out the four arithmetic operations and extracting square roots. Instead of the basis in Lemma 27, we select the maximal independent subset $B^* = \{N_\alpha\}$ from the set of all tetrahedra of the form $T(x, y)$ with $x, y \in R^*$ (this can be done without using the axiom of choice, since the set R^* is *countable*). Moreover, it is sufficient to prove Lemma 24 in the case where the function $F(x, y)$ is defined only for $x, y \in \,]0, 1[\, \cap R^*$. Then $\Phi(u, v)$ is defined for all u, v in an additive group R', containing only a *countable* number of elements (the positive elements of R' are the numbers of the form $-\ln x$, where $x \in \,]0, 1[\, \cap R^*$). Instead of $R \times R$, we only consider the group $R' \times R$ (with the same addition rule (66)), and this allows us to complete the proof of Lemma 24 without using Zorn's lemma. The same applies to Lemmas 25 and 26 (with $x, y, \lambda \in R^*$).

Furthermore, in Lemma 28 we consider the functions $\varphi_\alpha(x)$ only on the set S consisting of all x for which $\sin x \in R^*$. It is easy to see that S is an additive group. In fact, if $x \in S$, i.e., if $\sin x \in R^*$, we conclude successively

$$\sin^2 x \in R^*, \quad \cos^2 x = 1 - \sin^2 x \in R^*,$$
$$|\cos x| = \sqrt{\cos^2 x} \in R^*, \quad \cos x = \pm |\cos x| \in R^*.$$

Therefore, if $x, y \in S$, we have

$$\sin x \in R^*, \quad \sin y \in R^*, \quad \cos x \in R^*, \quad \cos y \in R^*,$$

and hence $\sin(x + y) \in R^*$, i.e., $x + y \in S$. It is also clear that $-x \in S$ (if $x \in S$); i.e., S is a group under addition.

We note further that all the simplexes μT (x, y) figuring in the expansion (95), and in the analogous expansion for the polyhedron B, have the property and their dihedral angles belong to the set S (by virtue of the formulas $\sin^2 \gamma_1 = x$, $\sin^2 \gamma_2 = y$, and $\sin^2 (\pi/2 - \gamma_3) = xy$ obtained in proving (89)). This implies that the invariant (88) is defined for all tetrahedra μT (x, y), where $x, y \in R^*$, and hence for the polyhedra A and B (see (95)). Finally, formula (94) is an immediate consequence of (90) and (95) (and similarly for B), and this allows us to complete the proof of Theorem 25.

§18. Polyhedra equidecomposable with a cube

In 1896 Hill [34] gave the first examples of tetrahedra equidecomposable with a cube, and one of these was given above (see Fig. 59). By now, many other examples of such tetrahedra are known. This section is devoted to a survey of various results along these lines.

In principle, the Dehn–Sydler theorem gives a *necessary and sufficient* condition for equidecomposability of a tetrahedron A and a cube; in fact, all the Dehn invariants of A must equal zero. However, it is difficult to use this condition for *finding* such tetrahedra, and a much more suitable tool is given by the following theorem, due to Sydler [50]:

Theorem 26. *If a polyhedron equidecomposable with a cube can be decomposed into several congruent polyhedra, then each of these polyhedra is also equidecomposable with a cube.*

PROOF. Let $P = A_1 + \ldots + A_p$, where P is a polyhedron equidecomposable with a cube, and $A_1, \ldots, A_p$ are polyhedra each of which is congruent to A. In other words, let $P \sim p \cdot A$. By Lemma 16, we have $pA \sim p \cdot A + Q$, where $Q \in Z_2$. Therefore $pA \sim P + Q$, and hence, by Lemma 19, the polyhedron pA is equidecomposable with the cube of the same volume. But then the same is true of the polyhedron A homothetic to pA.

We now give some examples of the application of this theorem.

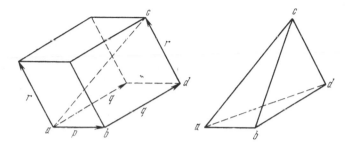

FIG. 89

Let P be the parallelepiped constructed on three vectors p, q, r of the same length which pairwise make the same angle ($p = \overrightarrow{ab}$, $q = \overrightarrow{bd}$, $r = \overrightarrow{dc}$ in Fig. 89). The three planes, each passing through the diagonal ac and two opposite vertices of the parallelepiped P, divide P into six *congruent* tetrahedra (cf. Fig. 86). By Theorem 26, each of these tetrahedra (for example, the tetrahedron $abcd$ in Fig. 89) is equidecomposable with a cube. These are *Hill's tetrahedra of the first type*. In particular, if P is a cube, we get the tetrahedron considered in §12 (see Fig. 59).

Hill tetrahedra of the first type can also be obtained in a different way. Let Π be the plane which goes through the point a and is perpendicular to the line ac. The orthogonal projections of the vectors p, q, r onto the plane Π are all of the *same* length, and hence the projection of the polygonal line $abdc$ onto the plane Π is an *equilateral* triangle T (Fig. 90). If l denotes the distance of the point b from the plane Π, then the point d is at distance $2l$ from Π, while the point c is at distance $3l$ from Π (since the vectors p, q, r all have the same projection onto the line ac). Thus the construction of a Hill tetrahedron of the first type can be described as follows: Consider a right prism whose perpendicular cross section is an equilateral triangle T with vertices a, m, n, and lay off segments mb, nd, ac of lengths l, $2l$, $3l$, respectively (where l is an arbitrary positive number) along the edges of the prism (on the same side of the plane of the triangle T). Then a, b, c, d are the vertices of a Hill tetrahedron of the first type.

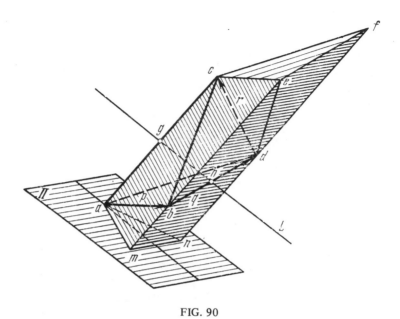

FIG. 90

In Fig. 90 we have laid off segments *be* and *df* of length $3l$ along the edges of the prism. The polyhedron Q with vertices *a, b, c, d, e, f* is an (oblique) triangular prism with base *abd*, and each of the tetrahedra *bcde, cdef* is congruent to *abcd* (the points *a, b, d, c, e* are carried into *b, d, c, e, f*, respectively, by the motion which is the composition of a rotation through the angle $2\pi/3$ about the axis of the prism and a translation by the vector $\frac{1}{3}(p + q + r) = \frac{1}{3}\overrightarrow{ac}$.

Thus *we can assemble the triangular prism Q from three tetrahedra congruent to a Hill tetrahedron of the first type.* Because of Theorem 26, this gives another proof of the fact that this tetrahedron is equidecomposable with a cube.

We now note that the line L going through the midpoints g and h of the edges *ac* and *bd* (see Fig. 90) is an axis of symmetry of the tetrahedron *abcd*. In fact, a rotation through the angle π about this axis carries the tetrahedron *abcd* into itself. Therefore *every plane through the line L divides the tetrahedron abcd into two congruent pieces.* In particular, suppose that in the tetrahedron *abcd* we draw the plane through the edge *bd* and the midpoint *g* of the edge *ac*. Then we get two congruent tetrahedra (Fig. 91), each equidecomposable with a cube, by Theorem 26. These are *Hill's tetrahedra of the second type.* Similarly, drawing the line through the edge *ac* and the midpoint *h* of the side *bd*, we again get two congruent tetrahedra (Fig. 92), this time called *Hill's tetrahedra of the third type,* each of which is equidecomposable with a cube. Let α be the dihedral angle at the edge *ab*. Then we can easily calculate the other dihedral angles of the Hill tetrahedra $H_1(\alpha)$, $H_2(\alpha)$ and $H_3(\alpha)$ (of the first, second and third type, respectively), as well as the lengths of their edges. This data is given in the table on pp. 170-173.

In addition to these *infinite families,* the table also gives some special tetrahedra equidecomposable with a cube. The tetrahedron T_0 was also found by Hill. Of course, Hill proved the equidecomposability of T_0 with a cube *directly* (by decomposing T_0 into pieces). From our standpoint, this fact is obvious, since the dihedral angles are rational multiples of the number π, so that all the Dehn invariants $f(T_0)$ equal zero, and T_0 is equidecomposable with a cube by the Dehn-Sydler theorem. It is easily verified that all the

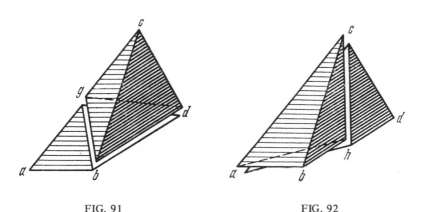

FIG. 91   FIG. 92

$H_1(\alpha)$, $H_2(\alpha)$, $H_3(\alpha)$

Edges	$H_1(\alpha)$ Lengths	Dih. angles	$H_2(\alpha)$ Lengths	Dih. angles	$H_3(\alpha)$ Lengths	Dih. angles
ab	$\sin\alpha$	α	$2\sin\alpha$	α	$2\sin\alpha$	α
ac	$\sqrt{3}\cos\alpha$	$\pi/3$	$\sqrt{3}\cos\alpha$	$\pi/3$	$\sqrt{12}\cos\alpha$	$\pi/6$
ad	1	$\pi/2$	2	$\pi/2$	$\sqrt{2+\sin^2\alpha}$	$\pi-\arccos\left(\dfrac{1}{\sqrt{3}}\cos\alpha\right)$
bc	1	$\pi/2$	$\sqrt{5\sin^2\alpha-1}$	$\pi-\arccos\left(\dfrac{1}{2}\cot\alpha\right)$	2	$\pi/2$
bd	$\sin\alpha$	$\pi-2\alpha$	$2\sin\alpha$	$\pi/2-\alpha$	$\sin\alpha$	$\pi-2\alpha$
cd	$\sin\alpha$	α	$\sqrt{5\sin^2\alpha-1}$	$\arccos\left(\dfrac{1}{2}\cot\alpha\right)$	$\sqrt{2+\sin^2\alpha}$	$\arccos\left(\dfrac{1}{\sqrt{3}}\cos\alpha\right)$

T_0, T_1, T_2, T_3

Edges	T_0 Lengths	Dih. angles	T_1 Lengths	Dih. angles	T_2 Lengths	Dih. angles	T_3 Lengths	Dih. angles
ab	$\sqrt{3}$	$\pi/3$	$\tau=(\sqrt{5}+1)/2$	$\pi/2$	$\sqrt{3}$	$2\pi/3$	$\sqrt[4]{5}\,\sqrt{\tau}$	$\pi/5$
ac	$\sqrt{2}$	$\pi/2$	$1/\tau$	$\pi/2$	$\sqrt[4]{5}\,\sqrt{\tau}$	$\pi/5$	$\sqrt{3}$	$\pi/3$
ad	2	$\pi/4$	1	$\pi/2$	$\sqrt[4]{5}\,\sqrt{\tau}$	$\pi/5$	2	$\pi/2$
bc	1	$\pi/2$	$\sqrt{3}$	$\pi/3$	$\sqrt[4]{5}/\sqrt{\tau}$	$2\pi/5$	$\sqrt[4]{5}/\sqrt{\tau}$	$3\pi/5$
bd	$\sqrt{3}$	$\pi/3$	$\sqrt[4]{5}\,\sqrt{\tau}$	$\pi/5$	$\sqrt[4]{5}/\sqrt{\tau}$	$2\pi/5$	$\sqrt{3}/\tau$	$\pi/3$
cd	$\sqrt{2}$	$\pi/2$	$\sqrt[4]{5}/\sqrt{\tau}$	$2\pi/5$	2	$\pi/2$	$\sqrt[4]{5}/\sqrt{\tau^3}$	$2\pi/5$

Edges	T_4 Lengths	Dih. angles	T_5 Lengths	Dih. angles	T_6 Lengths	Dih. angles	T_7 Lengths	Dih. angles
ab	$\sqrt[4]{5}\,\sqrt{\tau^3}$	$\pi/5$	$\sqrt[4]{5}\,\sqrt{\tau}$	$2\pi/5$	$\sqrt[4]{5}/\sqrt{\tau}$	$4\pi/5$	$\sqrt[4]{5}\,\sqrt{\tau}$	$3\pi/5$
ac	$\sqrt{3}\tau$	$\pi/3$	$\sqrt{3}$	$\pi/3$	$\sqrt[4]{5}\,\sqrt{\tau}$	$\pi/5$	$\sqrt{3}$	$\pi/3$
ad	2	$\pi/2$	$\sqrt{3}$	$\pi/3$	$\sqrt[4]{5}\,\sqrt{\tau}$	$\pi/5$	$\sqrt[4]{5}/\sqrt{\tau}$	$\pi/5$
bc	$\sqrt[4]{5}\,\sqrt{\tau}$	$\pi/5$	$\sqrt[4]{5}/\sqrt{\tau}$	$2\pi/5$	$\sqrt{3}$	$\pi/3$	$\sqrt[4]{5}/\sqrt{\tau}$	$\pi/5$
bd	$\sqrt{3}$	$2\pi/3$	$\sqrt[4]{5}/\sqrt{\tau}$	$2\pi/5$	$\sqrt{3}$	$\pi/3$	$\sqrt{3}$	$\pi/3$
cd	$\sqrt[4]{5}/\sqrt{\tau}$	$3\pi/5$	$2/\tau$	$\pi/2$	2τ	$\pi/2$	$\sqrt[4]{5}\,\sqrt{\tau}$	$3\pi/5$

Edges	T_8 Lengths	Dih. angles	T_9 Lengths	Dih. angles	T_{10} Lengths	Dih. angles
ab	$\sqrt[4]{5}\,\sqrt{\tau}$	$3\pi/5$	$\sqrt[4]{5}\,\sqrt{\tau}$	$3\pi/5$	$\sqrt[4]{5}\,\sqrt{\tau}$	$3\pi/10$
ac	$\sqrt{3}$	$\pi/6$	$\sqrt{3}$	$\pi/3$	$\sqrt{3}$	$\pi/3$
ad	$\sqrt{7/2}$	$\alpha_1 = \arctan\sqrt{7/5}$	$\sqrt[4]{5}/(2\sqrt{\tau})$	$\pi/5$	$\sqrt{7-3/\tau}/2$	$\alpha_3 = \arctan\sqrt{9+2\sqrt{5}}$
bc	$\sqrt[4]{5}/\sqrt{\tau}$	$\pi/5$	$\sqrt[4]{5}/\sqrt{\tau}$	$\pi/10$	$\sqrt[4]{5}/\sqrt{\tau}$	$\pi/5$
bd	$\sqrt{3/2}$	$\pi/3$	$\sqrt{7+3\tau}/2$	$\alpha_2 = \arctan\sqrt{9-2\sqrt{5}}$	$\sqrt{7-3/\tau}/2$	$\pi - \alpha_3$
cd	$\sqrt{7/2}$	$\pi - \alpha_1$	$\sqrt{7+3\tau}/2$	$\pi - \alpha_2$	$\sqrt[4]{5}\,\sqrt{\tau}/2$	$3\pi/5$

Edges	T_{11} Lengths	T_{11} Dih. angles	T_{12} Lengths	T_{12} Dih. angles	T_{13} Lengths	T_{13} Dih. angles	T_{14} Lengths	T_{14} Dih. angles
ab	$\sqrt[4]{5}\sqrt{\tau}$	$3\pi/10$	$\sqrt{3}$	$\pi/6$	$\sqrt{2+\tau}$	$\pi/5$	$\sqrt{2+\tau}$	$\pi/5$
ac	$\sqrt{3}$	$\pi/6$	$\sqrt{3}$	$\pi/6$	$\sqrt{6-3\tau}$	$\pi/3$	$2\sqrt{6-3\tau}$	$\pi/3$
ad	1	$\alpha_4 = \pi - \arctan 2\tau^2$	$\sqrt{5}/2$	$\alpha_7 = \pi - \arccos{}^2/_3$	$2/\tau$	$\pi/2$	$2\sqrt{5}-2$	$\pi/2$
bc	$\sqrt[4]{5}/\sqrt{\tau}$	$\pi/10$	2	$\pi/4$	$2/\tau$	$\pi/2$	$\sqrt{7+3\tau/\tau^2}$	$\pi-\alpha_2$
bd	1	$\alpha_5 = \pi - \arctan 2$	$\sqrt{5}/2$	$\pi - \alpha_7/2$	$\sqrt{6-3\tau}$	$\pi/3$	$\sqrt{7+3\tau/\tau^2}$	α_2
cd	1	$\alpha_6 = 2\pi - \alpha_4 - \alpha_5$	$\sqrt{5}/2$	$\pi - \alpha_7/2$	$\sqrt{18-11\tau}$	$3\pi/5$	$2\sqrt{18-11\tau}$	$3\pi/10$

Edges	T_{15} Lengths	T_{15} Dih. angles	T_{16} Lengths	T_{16} Dih. angles	T_{17} Lengths	T_{17} Dih. angles	T_{18} Lengths	T_{18} Dih. angles
ab	$2\sqrt{2+\tau}$	$\pi/10$	$\sqrt{2+\tau}$	$\pi/5$	$\sqrt{2+\tau}$	$\pi/5$	$\sqrt{7-4\tau}$	$\pi/5$
ac	$2\sqrt{6-3\tau}$	$\pi/3$	$\tau\sqrt{3}$	$\pi/3$	$\tau\sqrt{3}$	$\pi/3$	$\sqrt{3-\tau}$	$\pi/5$
ad	$\sqrt{10-3\tau}$	$\pi - \alpha_3$	2τ	$\pi/2$	2	$\pi/2$	$\sqrt{6-3\tau}$	$2\pi/3$
bc	$2\sqrt{5}-2$	$\pi/2$	$\sqrt{3}$	$2\pi/3$	$\sqrt{3}$	$\pi/3$	$\sqrt{6-3\tau}$	$2\pi/3$
bd	$\sqrt{10-3\tau}$	α_3	$\sqrt{2+\tau}$	$2\pi/5$	$\sqrt{6-3\tau}$	$2\pi/3$	$\sqrt{3}/\tau^2$	$\pi/3$
cd	$\sqrt{18-11\tau}$	$3\pi/5$	$\sqrt{3-\tau}$	$\pi/5$	$\sqrt{3-\tau}$	$2\pi/5$	$\sqrt{7-4\tau}$	$\pi/5$

Edges	T_{19} Lengths	T_{19} Dih. angles	T_{20} Lengths	T_{20} Dih. angles	T_{21} Lengths	T_{21} Dih. angles	T_{22} Lengths	T_{22} Dih. angles
ab	$2\sqrt{3}(\tau-1)$	$\pi/6$	$2\sqrt{2+\tau}$	$\pi/10$	$\sqrt{3-\tau}$	$\pi/5$	$\tau\sqrt{3}$	$\pi/3$
ac	$2\sqrt{3-\tau}$	$\pi/5$	$2\sqrt{3-\tau}$	$\pi/5$	$\sqrt{3}$	$\pi/3$	$2\sqrt{3}$	$\pi/3$
ad	$\sqrt{10-3\tau}$	$\pi-\alpha_8$	$\sqrt{6+\tau}$	$\pi-\alpha_9$	$\sqrt{3-\tau}$	$3\pi/5$	$2\sqrt{2+\tau}$	$2\pi/5$
bc	$2\sqrt{3}$	$2\pi/3$	$2\sqrt{3}$	$2\pi/3$	$\sqrt{2+\tau}$	$2\pi/5$	$\sqrt{7-\tau}$	$\pi-\alpha_{10}$
bd	$\sqrt{10-3\tau}$	$\alpha_8=\arccos u,$ $u=\tau^2/(2\sqrt{3})$	$\sqrt{6+\tau}$	$\alpha_9=\arccos v,$ $v=\tau^{3/2}/(2\sqrt[4]{5})$	$\sqrt{5}-1$	$\pi/2$	$\sqrt{7-\tau}$	$\alpha_{10}=\arctan w,$ $w=\sqrt{11+16\tau}$
cd	$\sqrt{2+\tau}$	$\pi/5$	$\sqrt{3}(\tau-1)$	$\pi/3$	$\sqrt{3}(\tau-1)$	$\pi/3$	$2\sqrt{3-\tau}$	$3\pi/10$

Edges	T_{23} Lengths	T_{23} Dih. angles	T_{24} Lengths	T_{24} Dih. angles	T_{25} Lengths	T_{25} Dih. angles	T_{26} Lengths	T_{26} Dih. angles
ab	$\sqrt{3}$	$\pi/3$	$\sqrt{3}$	$\pi/6$	$\sqrt{6-3\tau}$	α_{12}	$\sqrt{3}$	$\pi/3-\alpha_{12}$
ac	$\sqrt{6-3\tau}$	$\pi/3$	$\sqrt{6-3\tau}$	$\pi/3$	$\sqrt{3}$	$\pi/3$	$\sqrt{3}$	$\alpha_{12}=\arctan\sqrt{3/5}$
ad	$\sqrt{3-\tau}$	$2\pi/5$	$\sqrt{11-4\tau}/2$	$\pi-\alpha_{11}$	$2\sqrt{4-2\tau}$	$\pi/2$	$\sqrt{3-\tau}$	$4\pi/5$
bc	$\sqrt{3-\tau}$	$2\pi/5$	$\sqrt{3-\tau}$	$2\pi/5$	$\sqrt{3-\tau}$	$3\pi/5$	$2\sqrt{2}$	$\pi/2$
bd	$\sqrt{6-3\tau}$	$\pi/3$	$\sqrt{11-4\tau}/2$	$\alpha_{11}=\arccos z,$ $z=1/(2\sqrt{3}\tau^2)$	$\sqrt{6-3\tau}$	$2\pi/3-\alpha_{12}$	$\sqrt{2+\tau}$	$2\pi/5$
cd	$\sqrt{7-4\tau}$	$3\pi/5$	$\sqrt{7-4\tau}/2$	$3\pi/5$	$\sqrt{7-4\tau}$	$\pi/5$	$\sqrt{3}(\tau-1)$	$\pi/3$

173

Dehn invariants also equal zero for the rest of the tetrahedra listed in the table.

The fact that each of the tetrahedra $T_1 - T_4$ is equidecomposable with a cube was proved in Sydler's paper [52], by using Theorem 26. The tetrahedra T_5 and T_6 were found in 1958 by Goldberg [14], and in 1962 Lenhardt [43] added five more tetrahedra $T_7 - T_{11}$ to the list. We note that the tetrahedron T_8, for example, has edges (ad and cd) at which the dihedral angles are not rational multiples of π. However, in calculating any Dehn invariant $f(T_8)$, these edges contribute zero:

$$\frac{\sqrt{7}}{2} f(\alpha_1) + \frac{\sqrt{7}}{2} f(\pi - \alpha_1) =$$

$$= \frac{\sqrt{7}}{2} f(\alpha_1) + \frac{\sqrt{7}}{2} f(\pi) - \frac{\sqrt{7}}{2} f(\alpha_1) = 0.$$

Since the remaining dihedral angles are rational multiples of π, we have $f(T_8) = 0$, and hence, by Theorem 25, the tetrahedron T_8 is equidecomposable with a cube. Lenhardt's tetrahedra are distinguished by the fact that none of them contains a dihedral angle equal to a right angle.

The tetrahedron T_{12} was found back in 1923 by Sommerville in a paper [48] dealing with Hilbert's *eighteenth* problem. The fact that this tetrahedron is equidecomposable with a cube was noted in 1969 by Goldberg [15]. Finally, the remaining tetrahedra $T_{13} - T_{26}$ come from a paper by Goldberg [17], where the "tricks of the trade" are also revealed. Following this paper, we now indicate ways of finding tetrahedra equidecomposable with a cube.

Figure 93 shows the Sydler tetrahedron T_4, which is divided into two smaller tetrahedra by the plane abe perpendicular to the face acd. Each edge is labelled with the corresponding dihedral angle. It is easily verified (by examining dihedral angles) that $abce$ is the Sydler tetrahedron T_3, whereas $abde$ is a *new* tetrahedron, which we denote by T_{13}. Thus $T_4 = T_3 + T_{13}$, and since T_3 and T_4 are equidecomposable with a cube, so is T_{13}.

The tetrahedron T_{13} has an axis of symmetry going through the midpoints of two edges. Therefore, just as in the case of Hill's

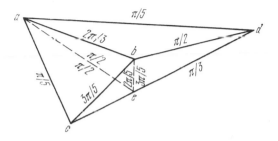

FIG. 93

tetrahedron of the first type, it can be divided into two congruent tetrahedra, and this gives T_{14}. Another way of dividing the tetrahedron T_{13} into two congruent tetrahedra gives T_{15}. Moreover $T_{14} = T_9 + T_{16}$, and so on.

It is interesting to note that for each of the tetrahedra T_k, $k = 0, \ldots, 26$, Bricard's condition (30) holds in a more specialized form, i.e., all the coefficients n_i are equal, so that the *sum* Σ of all six dihedral angles is a rational multiple of π. The same is true for the Hill tetrahedra of the first and second type.

Besides tetrahedra, there are other polyhedra known to be equidecomposable with a cube, for example, the *orthogonal polyhedron*[22] given in Sydler's paper [51]. Every dihedral angle of an orthogonal polyhedron equals $\pi/2$ or $3\pi/2$, and hence, by Theorem 25, every such polyhedron is equidecomposable with a cube. A number of interesting examples of centrally symmetric orthogonal (nonconvex) polyhedra can be found in Jessen's paper [35]; each of them is of the same topological type as the *regular icosahedron*. One of Jessen's icosahedra is shown in Fig. 94. Its vertices are at the points $(\pm 2, \pm 1, 0)$ $(0, \pm 2, \pm 1)$, $(\pm 1, 0, \pm 2)$ in a rectangular coordinate system; eight faces are equilateral triangles, and the other twelve are

---

[22] A polyhedron is said to be *orthogonal* if two faces are perpendicular whenever they have a common edge. Orthogonal polyhedra play an important role in Sydler's original treatment [54] of Theorem 25.

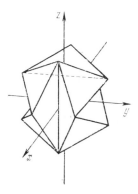

FIG. 94

isosceles triangles. The convex hull of Jessen's icosahedron (Fig. 94) resembles a regular icosahedron.

In conclusion, we say a few words about the connection between our subject and Hilbert's *eighteenth* problem. This is the problem of finding polyhedra such that all of space can be filled by juxtaposition (without overlapping) of congruent copies of the polyhedra (see [33], pp. 466–467). Let G be a group of motions of the space R^3, and suppose R^3 can be completely filled by juxtaposition of G-congruent copies of a polyhedron. Then this filling of R^3 will be called a *G-tiling*. Every triangle gives a tiling in the plane (Fig. 95), but not every tetrahedron gives a tiling in R^3.

Four tetrahedra giving D_0-tilings of space were found in 1923 by Sommerville [48], [49]. In the notation adopted above, these are the tetrahedra H_1 $(\pi/3)$, T_0, H_2 $(\pi/4)$ and T_{12}. Three of them were discovered independently by Davies [11]. Somewhat later Baumgartner [3], [4] found four tetrahedra giving D_0-tilings, of which three were Somerville tetrahedra, while one (namely, H_2 $(\pi/3)$) was new.

As we have seen (cf. Fig. 89), six congruent Hill tetrahedra H_1 (α) fill a parallelepiped. Thus, since a parallelepiped gives a tiling of space, the same must be true of the Hill tetrahedron H_1 (α). In general, this method gives a D-tiling, and not a D_0-tiling, since three

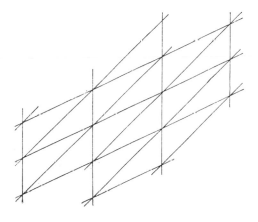

FIG. 95

of the six tetrahedra are obtained from the tetrahedron *abcd* by making motions that reverse the orientation. Evidently, it was just this fact that motivated investigators in their attempts to find *special cases* of Hill tetrahedra that give D_0-tilings.

In 1974 Goldberg [18] noted that *all* Hill tetrahedra give D_0-tilings, and his argument can be presented as follows: Figure 90 shows that three D_0-congruent Hill tetrahedra H_1 (α) can be put together to form an oblique triangular prism, and congruent copies of this oblique prism can then be used to assemble an infinite prism whose perpendicular cross section is an equilateral triangle. From two such prisms we can assemble an infinite prism whose perpendicular cross section is a rhombus, and we can then use congruent copies of this rhombic "beam" to fill all of R^3. Moreover, since two D_0-congruent tetrahedra H_2 (α) can be assembled to form the tetrahedron H_1 (α), and the same can be done with two D_0-congruent tetrahedra H_3 (α), it follows that H_2 (α) and H_3 (α) also give D_0-tilings.

Thus H_1 (α), H_2 (α), H_3 (α), T_0 and T_{12} all give D_0-tilings. The existence of the last two tilings follows from the formulas $2 \cdot T_0 = H_2$ ($\pi/4$) and $4 \cdot T_{12} = H_1$ ($\pi/3$). (If *abcd* is the face of a cube and q is the center of the face, while o is the center of the cube itself, then $oabc$ is $H_2(\pi/4)$ and each of the tetrahedra $aboq, bcoq$ is T_0; moreover, if m is the centroid of the tetrahedron H_1 $(\pi/3)$, then each of the four tetrahedra with common vertex m, whose bases are the faces of H_1 $(\pi/3)$, is T_{12}.) No other tetrahedra giving D_0-tilings are known.

Examples of pentahedra giving D_0-tilings can be found in Goldberg's papers [16], [19]. Here we give just two examples of such pentahedra. We have already noted that the Hill tetrahedron H_1 (α) has an axis of symmetry L (Fig. 90). Any plane which goes through L and does not contain either of the edges ac, bd intersects H_1 (α) in two D_0-congruent pentahedra (Fig. 96), each of which gives a D_0-tiling. To construct the second example, we consider an infinite prism P, whose perpendicular cross section is a regular n-gon, where n is one of the numbers 3, 4, 6. Let f denote the composition of a rotation through the angle $2\pi/n$ about this axis and some translation along the axis, and let Π be a plane, not parallel to the axis of the prism, such that the n-gons $\Pi \cap P$ and $f(\Pi) \cap P$ have no points in common. Then the polyhedron Q, equal to the part of the prism P lying between these two n-gons (Fig. 97) gives a D_0-tiling. This follows from the fact that the polyhedra

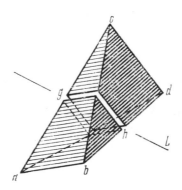

FIG. 96

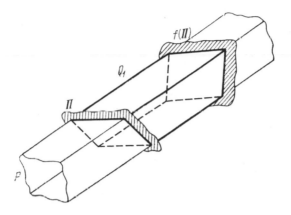

FIG. 97

$$Q_1, \; Q_2 = f\,(Q_1), \; Q_3 = f\,(Q_2), \; \ldots, Q_n = f\,(Q_{n-1})$$

make up an oblique prism, since an n-fold application of the motion f leads to a translation (by Theorem 26, the fact that the polyhedra $Q_1, \; \ldots, \; Q_n$ make up a prism also implies that Q_1 is equidecomposable with a cube). We can then use congruent copies of this oblique prism to fill a whole infinite prism P, and afterwards use congruent copies of P to fill all of R^3. For $n = 3$ the polyhedron Q_1 is a pentahedron giving D_0-tiling.

Baumgartner [3], [4] and Danzer [10] have investigated simplexes which give tilings in R^n, and Hadwiger has shown that the n-dimensional simplexes generalizing Hill's tetrahedra are equidecomposable with an n-dimensional cube.

§19. Equidecomposability of polyhedra with respect to the group of translations

In this section we discuss the necessary and sufficient condition for T-equidecomposability of polyhedra in R^3, found in 1968 by Hadwiger [29].

Let Π be a plane in R^3, and let p be a straight line contained in Π.

The line p divides Π into two half-planes, one of which we call the *positive* half-plane; moreover, the plane Π determines two half-spaces, one of which we call the *positive* half-space. A quadruple of this kind (consisting of a line, a plane, a half-plane and a half-space) will be called a 1-*rig*.

Similarly, let P be a plane in R^3, and let one of the two half-spaces determined by P be called the *positive* half-space. Then such a pair (consisting of a plane and a half-space) will be called a 2-*rig*.

Now let σ be a 1-rig, and let p, Π be the line and plane contained in σ. Moreover, let M be a polyhedron which has a face Γ parallel to the plane Π, where Γ has a side r parallel to the line p. If the face Γ adjoins r from the positive side (Fig. 98), we assign the edge r the coefficient $\varepsilon_1 = 1$, while if Γ adjoins r from the negative side, we assign r the coefficient $\varepsilon_1 = -1$. Moreover, if M adjoins the face Γ from the positive side (Fig. 98), we assign the face Γ the coefficient $\varepsilon_2 = 1$, while if M adjoins Γ from the negative side, we assign Γ the coefficient $\varepsilon_2 = -1$. By the *weight* of the *edge* r in the polyhedron M we mean the product $\varepsilon_1\varepsilon_2 l$, where l is the *length* of the edge r. However, if r is not parallel to the line p, or if $r \parallel p$ but M has no face which is parallel to the plane Π and adjoins the edge r, then the weight of r is taken to be zero. Let $K_\sigma(M)$ denote the sum of the weights of all the edges of the polyhedron M.

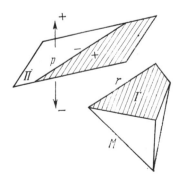

FIG. 98

Next let τ be a 2-rig, and let P be the plane contained in τ. Moreover, let M be a polyhedron which has a face Γ parallel to P. If M adjoins the face Γ from the positive side, we assign the face Γ the coefficient $\varepsilon_2 = 1$, while if M adjoins Γ from the negative side, we assign Γ the coefficient $\varepsilon_2 = -1$. By the *weight* of the *face* Γ in the polyhedron M we mean the product $\varepsilon_2 s$, where s is the *area* of the face Γ. Finally, let $K_\tau(M)$ denote the sum of the weights of all the faces of the polyhedron M parallel to the plane P (if there are no such faces, then $K_\tau(M) = 0$).

As we now show, the functions $K_\sigma(M)$ and $K_\tau(M)$ are both *additive T-invariants*. The T-invariance is obvious (the values of $K_\sigma(M)$ and $K_\tau(M)$ do not change if the polyhedron M is subjected to a translation). To prove the additivity, let $A = P_1 + P_2$, and divide the edges of the polyhedra A, P_1, P_2 into *links*, as in the proof of Lemma 11. Then the invariants $K_\sigma(A)$, $K_\sigma(P_1)$, $K_\sigma(P_2)$ can be calculated by summing over the *links*, rather than over the edges. Consider the sum Σ of the weights of all the links belonging to the two polyhedra P_1 and P_2. If a link m parallel to p, lies *inside* the polyhedron A, then either both polyhedra P_1 and P_2 have faces adjoining m and parallel to the plane Π, or neither does (see Fig. 99, which, like Fig. 65, represents a cross section by a plane perpendicular to m). In either case, the sum of the

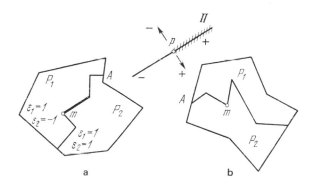

a   b

FIG. 99

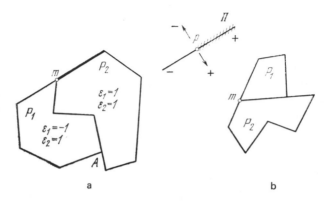

a   b

FIG. 100

weights of the link m equals zero. The same is true for links parallel to p which lie on faces of the polyhedron A (Fig. 100). Finally, if a link m lies on an edge of the polyhedron A (Fig. 101), then it contributes a term to the sum Σ equal to the weight of the link m in the polyhedron A. All this implies the validity of the formula

$$K_\sigma(A) = K_\sigma(P_1) + K_\sigma(P_2),$$

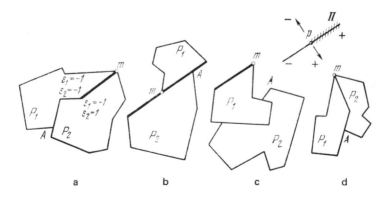

a   b   c   d

FIG. 101

thereby proving the additivity of the function K.

To prove the additivity of the function K_τ, let $A = P_1 + P_2$ as before, and consider all possible *cells*, by which we mean all possible *two-dimensional* intersections $\Gamma \cap \Gamma_1$, $\Gamma \cap \Gamma_2$, $\Gamma_1 \cap \Gamma_2$ of the faces Γ, Γ_1, Γ_2 of the polyhedra A, P_1, P_2. Every face of each polyhedron A, P_1, P_2 is the union of a number of cells with pairwise disjoint interiors, and to calculate $K_\tau(A)$, $K_\tau(P_1)$, $K_\tau(P_2)$, we can take the sum of the areas of the *cells* (with appropriate coefficients), rather than the sum of the areas of the faces. If a cell γ, parallel to the plane P, lies entirely inside A (with the possible exception of boundary points), then *both* polyhedra P_1 and P_2 adjoin γ, but from different sides. Hence the area of γ enters $K_\tau(P_1)$ with one sign and $K_\tau(P_2)$ with the opposite sign, so that the two terms contribute zero to the sum $K_\tau(P_1) + K_\tau(P_2)$. On the other hand, if γ is a cell lying on the boundary of A, then only *one* of the polyhedra P_1, P_2 adjoins γ, and it does so from the same side as A itself. Hence, in this case, the area of γ enters the sum $K_\tau(P_1) + K_\tau(P_2)$ with the same sign as it enters $K_\tau(A)$. Thus, finally,

$$K_\tau(A) = K_\tau(P_1) + K_\tau(P_2),$$

and the additivity of K_τ is proved.

Lemma 30 (see [37]). *Let M be a polygon in R^3, and let I be a line segment which is not parallel to the plane of M. Then, for every $\lambda > 0$, there exist polyhedra P, $Q \in Z_3$ such that*

$$I \times \lambda M + P \underset{T}{\sim} \lambda I \times M + Q.$$

PROOF. First let M be a triangle with vertices a, b, c, and let d be the midpoint of the side bc. Setting $e_2 = \overrightarrow{cd}$, $e_3 = \overrightarrow{da}$, we let e_1 denote the interval I, on which some direction is chosen. Then (Fig. 102)

$$M = [e_2, e_3] + [-e_2, e_3]. \tag{96}$$

According to Lemma 14,

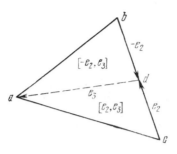

FIG. 102

$$(1 + \lambda) [e_1, e_2, e_3] \underset{T}{\sim} [e_1, e_2, e_3] +$$
$$+ \lambda [e_1, e_2, e_3] + [e_1] \times \lambda [e_2, e_3] + [e_1, e_2] \times \lambda [e_3],$$
$$(\lambda + 1) [e_1, e_2, e_3] \underset{T}{\sim} \lambda [e_1, e_2, e_3] +$$
$$+ [e_1, e_2, e_3] + \lambda [e_1] \times [e_2, e_3] + \lambda [e_1, e_2] \times [e_3].$$

Taking the last two terms in each of these formulas, we get polyhedra which are T-equicomplementable, and hence T-equidecomposable as well (Theorem 22):

$$[e_1] \times \lambda [e_2, e_3] + [e_1, e_2] \times \lambda [e_3] \underset{T}{\sim}$$
$$\underset{T}{\sim} \lambda [e_1] \times [e_2, e_3] + \lambda [e_1, e_2] \times [e_3]. \qquad (97)$$

Replacing e_1, e_2 by $-e_1, -e_2$, we write the analogous formula

$$[-e_1] \times \lambda [-e_2, e_3] + [-e_1, -e_2] \times \lambda [e_3] \underset{T}{\sim}$$
$$\underset{T}{\sim} \lambda [-e_1] \times [-e_2, e_3] + \lambda [-e_1, -e_2] \times [e_3]. \qquad (98)$$

But

$$[e_1, e_2] + [-e_1, -e_2] \underset{T}{\sim} W,$$

where W is a parallelogram, and each of the one-dimensional simplexes $[e_1]$, $[-e_1]$ coincides with the segment I. Therefore, adding (97) and (98), we obtain

$$I \times \lambda M + W \times \lambda\,[e_3] \underset{T}{\sim} \lambda I \times M + \lambda W \times [e_3]$$

(see (96)). Since $W \times \lambda\,[e_3]$ and $\lambda W \times [e_3]$ are parallelepipeds, the lemma holds in the case where M is a triangle. From this we deduce its validity for an arbitrary polygon M (by decomposing M into triangles).

Now let Q be a plane in R^3, and let π denote the operation of orthogonal projection onto the plane of Q. If M is a polygon that does not intersect the plane Q, where the plane of M is neither parallel nor perpendicular to Q, then by $W\,(M)$ we mean the convex hull of the set $M \bigcup \pi\,(M)$ (Fig. 103).

Lemma 31. *Let M_1, . . ., M_k, N_1, . . ., N_l be polygons, all lying on one side of a plane Q, whose planes are neither parallel nor perpendicular to Q and whose projections $\pi\,(M_1)$, . . ., $\pi\,(M_k)$,*

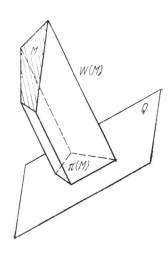

FIG. 103

$\pi\ (N_1),\ \ldots,\ \pi\ (N_l)$ *are pairwise disjoint. Suppose that*

$$K_\sigma\ (W\ (M_1)) + \ldots + K_\sigma\ (W(M_h)) =$$
$$= K_\sigma\ (W\ (N_1)) + \ldots + K_\sigma\ (W\ (N_l)), \quad (99)$$

$$K_\tau\ (W\ (M_1)) + \ldots + K_\tau\ (W\ (M_h)) =$$
$$= K_\tau\ (W\ (N_1)) + \ldots + K_\tau\ (W\ (N_l)), \quad (100)$$

$$v\ (W\ (M_1)) + \ldots + v\ (W\ (M_h)) =$$
$$= v\ (W\ (N_1)) + \ldots + v\ (W\ (N_l)) \quad (101)$$

(for arbitrary σ, τ*). Then* $W\ (M_1) + \ldots + W\ (M_h)$ *and* $W\ (N_1) + \ldots + W\ (N_l)$ *are T-equidecomposable.*

PROOF. We will assume that the planes of the polygons $M_1, \ldots, M_r, N_1, \ldots, N_q$ $(r \leqslant k, q \leqslant l)$ are parallel to a single plane P, while the planes of the remaining polygons are not parallel to P. Since the plane P is neither parallel nor perpendicular to the plane Q, we can assume, by making translations, if necessary (Fig. 104), that $M_1, \ldots, M_r, N_1, \ldots, N_q$ lie in the plane P.

Let τ^* denote the 2-rig specified by the plane P, with the positive half-space chosen in such a way that $W\ (M_1), \ldots, W\ (M_r)$, $W\ (N_1), \ldots, W\ (N_q)$ adjoin the faces $M_1, \ldots, M_r, N_1, \ldots, N_q$ from the positive side. Then

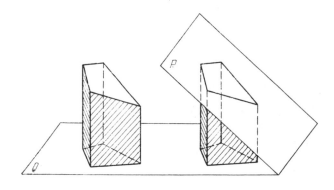

FIG. 104

$$K_{\tau^*}(W(M_i)) = s(M_i) \quad \text{if} \quad i = 1, \ldots, r;$$
$$K_{\tau^*}(W(M_j)) = 0 \quad \text{if} \quad j = r + 1, \ldots, k;$$
$$K_{\tau^*}(W(N_i)) = s(N_i) \quad \text{if} \quad i = 1, \ldots, q;$$
$$K_{\tau^*}(W(N_j)) = 0 \quad \text{if} \quad j = q + 1, \ldots, l,$$

and therefore, because of (100),

$$s(M_1) + \ldots + s(M_r) = s(N_1) + \ldots + s(N_q). \tag{102}$$

Moreover, let p be a line lying in the plane P, and let σ^* denote the 1-rig specified by the plane P and the line p (with the same choice of the positive half-space and some choice of the positive half-plane). Then

$$K_{\sigma^*}(W(M_i)) = J_p(M_i) \quad \text{if} \quad i = 1, \ldots, r;$$
$$K_{\sigma^*}(W(M_j)) = 0 \quad \text{if} \quad j = r + 1, \ldots, k;$$
$$K_{\sigma^*}(W(N_i)) = J_p(N_i) \quad \text{if} \quad i = 1, \ldots, q;$$
$$K_{\sigma^*}(W(N_j)) = 0 \quad \text{if} \quad j = q + 1, \ldots, l,$$

where J_p is the additive invariant (in the plane P) considered in §10. Therefore, because of (99),

$$J_p(M_1) + \ldots + J_p(M_r) = J_p(N_1) + \ldots + J_p(N_q). \tag{103}$$

Equation (103) is valid for every line $p \subset P$, and hence, by (102), the polygons $M_1 + \ldots + M_r$ and $N_1 + \ldots + N_q$, lying in the plane P, are T-equidecomposable (Theorem 16). In other words,

$$M_1 + \ldots + M_r = M_1^* + \ldots + M_s^*,$$
$$N_1 + \ldots + N_q = N_1^* + \ldots + N_s^*, \tag{104}$$

where M_i^* and N_i^* are T-congruent ($i = 1, \ldots, s$). It follows from (104) that

$$W(M_1) + \ldots + W(M_r) = W(M_1^*) + \ldots + W(M_s^*),$$
$$W(N_1) + \ldots + W(N_q) = W(N_1^*) + \ldots + W(N_s^*). \tag{105}$$

Similarly, if the planes of the polygons $M_{r+1}, \ldots, M_{r'}, N_{q+1}, \ldots,$ $N_{q'}$ are parallel to a single plane P', while the remaining polygons lie in planes not parallel to P', then

$$W(M_{r+1}) + \ldots + W(M_{r'}) = W(M^*_{s+1}) + \ldots + W(M^*_{s'}),$$
$$W(N_{q+1}) + \ldots + W(N_{q'}) = W(N^*_{s+1}) + \ldots + W(N^*_{s'}),$$
$$\tag{106}$$

where M^*_i and N^*_i are T-congruent $(i = s + 1, \ldots, s')$. Continuing in this way and adding the resulting equations (105), (106), $\ldots$, we finally obtain

$$W(M_1) + \ldots + W(M_k) = W(M^*_1) + \ldots + W(M^*_m),$$
$$W(N_1) + \ldots + W(N_l) = W(N^*_1) + \ldots + W(N^*_m),$$
$$\tag{107}$$

where M^*_i and N^*_i are T-congruent $(i = 1, \ldots, m)$.

Now let t_i denote the translation carrying M^*_i into N^*_i. For every $i = 1, \ldots, m$, we have either the formulas

$$W(M^*_i) = A_i, \; W(N^*_i) = B_i + \Pi; \; B_i = t_i(A_i)$$

(Fig. 105a) or the formulas

$$W(M^*_i) = A_i + \Pi, \; W(N^*_i) = B_i; \; B_i = t_i(A_i)$$

(Fig. 105b), where Π is a right prism whose bases are parallel to the plane Q. Thus

$$W(M^*_1) + \ldots + W(M^*_m) =$$
$$= A_1 + \ldots + A_m + \Pi_1 + \ldots + \Pi_\alpha,$$
$$W(N^*_1) + \ldots + W(N^*_m) =$$
$$\tag{108}$$
$$= B_1 + \ldots + B_m + \Pi'_1 + \ldots + \Pi'_\beta,$$

where Π_i, Π'_i are right prisms whose bases are parallel to Q.

The prism Π_i is of the form $\Pi_i = \lambda_i I \times L_i$, where $L_i \subset Q$ is some polygon and I is a line segment of length 1 perpendicular to Q.

By Lemma 30,

$$\Pi_i + P_i' \underset{T}{\sim} I \times \lambda_i L_i + P_i'', \quad P_i', \ P_i'' \in Z_3.$$

Moreover, by Lemma 18, $P_i' \underset{T}{\sim} I \times L_i'$, $P_i'' \underset{T}{\sim} I \times L_i''$, where L_i', $L_i'' \subset Q$ are certain rectangles. Therefore

$$\Pi_i + I \times L_i' \underset{T}{\sim} I \times \lambda_i L_i + I \times L_i'' = I \times (\lambda_i L_i + L_i'').$$

But $s(L_i') < s(\lambda_i L_i + L_i'')$, since the polyhedra in the left and right-hand sides have the same volume, and hence $\lambda_i L_i + L_i'' = L_i^* + L_i^{**}$, where $L_i' \underset{T}{\sim} L_i^*$ (Lemma 12). We now obtain

$$\Pi_i + I \times L_i' \underset{T}{\sim} I \times L_i^* + I \times L_i^{**}.$$

Since $I \times L_i' \underset{T}{\sim} I \times L_i^*$, the polyhedra Π_i and $I \times L_i^{**}$ are T-equicomplementable, and hence also T-equidecomposable (Theorem 22), i.e., $\Pi_i \underset{T}{\sim} I \times L_i^{**}$. Therefore, by (108),

$$W(M_1^*) + \ldots + W(M_m^*) \underset{T}{\sim} A_1 + \ldots A_m + I \times L,$$
$$L \subset Q, \quad (109)$$

where $L = L_1^{**} + \ldots + L_\alpha^{**}$, and similarly,

$$W(N_1^*) + \ldots + W(N_m^*) \underset{T}{\sim} B_1 + \ldots + B_m + I \times L',$$
$$L' \subset Q. \quad (110)$$

Comparing (101), (107), (109) and (110), we find that $s(L) = s(L')$.

Finally, let $p' \subset Q$ be any line, and let P' be the plane containing p' and perpendicular to the plane Q. Choosing one of the two half-spaces determined by the plane P' as the positive half-space, we denote the resulting 2-rig by τ'. It is easy to see that

$$K_{\tau'}(I \times L) = J_{p'}(L), \quad K_{\tau'}(I \times L') = J_{p'}(L'),$$

lower faces. Then (Fig. 107)

$$A + W(\Gamma_1^*) + \ldots + W(\Gamma_\beta^*) = W(\Gamma_1) + \ldots + W(\Gamma_\alpha).$$

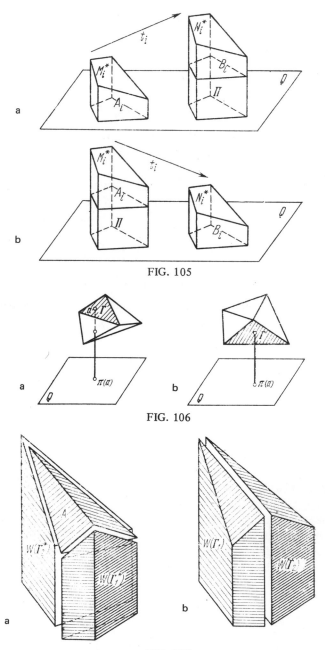

FIG. 105

FIG. 106

FIG. 107

Similarly,

$$R + W(\Lambda_1^*) + \ldots + W(\Delta_\delta^*) = W(\Delta_1) + \ldots + W(\Delta_\gamma),$$

where $J_{p'}$ is the additive invariant (in the plane Q) considered in §10. Thus it follows from (100), (107), (109) and (110) that

$$J_{p'}(L) = J_{p'}(L').$$

This formula holds for every line $p' \subset Q$, so that $L \underset{T}{\sim} L'$, by Theorem 16 (since $s(L) = s(L')$). Therefore $I \times L \underset{T}{\sim} I \times L'$, and hence, by (107), (109), (110) and the relations $A_i \underset{T}{\cong} B_i$, we have

$$W(M_1) + \ldots + W(M_k) \underset{T}{\sim} W(N_1) + \ldots + W(N_l).$$

thereby proving the lemma.

Theorem 27. *A necessary and sufficient condition for two polyhedra A, $B \subset R^3$ of equal volume to be T-equidecomposable is that the equalities $K_\sigma(A) = K_\sigma(B)$ and $K_\tau(A) = K_\tau(B)$ hold for every 1-rig σ and every 2-rig τ.*

PROOF. The necessity is an immediate consequence of Theorem 15. To prove the sufficiency, we argue as follows: Suppose $v(A) = v(B)$ and $K_\sigma(A) = K_\sigma(B)$, $K_\tau(A) = K_\tau(B)$ for arbitrary σ, τ. We represent A and B in the form

$$A = A_1 + \ldots + A_\mu, \quad B = B_1 + \ldots + B_\nu,$$

where A_i and B_j are *convex* polyhedra. Then we choose a plane Q which is neither parallel nor perpendicular to any face of any of the polyhedra A_i, B_j. We can assume (by making translations, if necessary) that the polyhedra A_i, B_j all lie on one side of the plane Q and that their orthogonal projections $\pi(A_1), \ldots, \pi(A_\mu), \pi(B_1), \ldots, \pi(B_\nu)$ onto the plane Q are pairwise disjoint.

Let Γ be any face of one of the polyhedra A_i, and let a be an interior point of Γ. If the segment $[a, \pi(a)]$ intersects the interior of the polyhedron A_i, we call Γ an *upper* face (Fig. 106a), but if not (Fig. 106b) we call Γ a *lower* face. The upper and lower faces of the polyhedra B_j are defined similarly. Now let $\Gamma_1, \ldots, \Gamma_\alpha$ be all the upper faces of all the polyhedra $A_1, \ldots, A_\mu$, and $\Gamma_1^*, \ldots, \Gamma_\beta^*$ all the where $\Delta_1, \ldots, \Delta_\gamma$ are the upper faces, and $\Delta_1^*, \ldots, \Delta_\delta^*$ the lower faces of the polyhedra $B_1, \ldots, B_\nu$. From these relations we get

$$A + \sum W(\Gamma_i^*) + \sum W(\Delta_i) = B + \sum W(\Delta_i^*) + \sum W(\Gamma_i),$$
$$\sum K_\sigma(W(\Gamma_i^*)) + \sum K_\sigma(W(\Delta_i)) = \sum K_\sigma(W(\Delta_i^*)) + \sum K_\sigma(W(\Gamma_i)),$$
$$\sum K_\tau(W(\Gamma_i^*)) + \sum K_\tau(W(\Delta_i)) = \sum K_\tau(W(\Delta_i^*)) + \sum K_\tau(W(\Gamma_i)),$$
$$\sum v(W(\Gamma_i^*)) + \sum v(W(\Delta_i)) = \sum v(W(\Delta_i^*)) + \sum v(W(\Gamma_i)). \quad (111)$$

By Lemma 31, it follows from the last three equations (valid for arbitrary rigs σ, τ) that

$$\sum W(\Gamma_i^*) + \sum W(\Delta_i) \underset{T}{\sim} \sum W(\Delta_i^*) + \sum W(\Gamma_i).$$

Therefore, by (111), the polyhedra A and B are T-equicomplementable, and hence (by Theorem 22) T-equidecomposable.

In particular, Theorem 27 implies the following interesting result, due to Hadwiger [29]: *A convex polyhedron A is T-equidecomposable with a cube of the same volume if and only if every face of A is a centrally symmetric polygon.* This result can be given a more elegant form by using a theorem due to Aleksandrov [1]. To this end, we note that there are three possibilities for the Minkowski sum of several line segments: 1) a line segment (if all the segments being added are parallel to a single line); 2) a convex centrally symmetric polygon (if the segments being added are parallel to a single plane but not to any one line); 3) a convex polyhedron (if the segments are not parallel to any one plane). A convex polyhedron which can be represented as the Minkowski sum of several line segments is called a *zonohedron.* For example, a prism whose base is a convex centrally symmetric polygon is a zonohedron. Aleksandrov's theorem (or, more exactly, its corollary) asserts that *a convex polyhedron is a zonohedron if and only if its faces are centrally symmetric.* For example, the Archimedean solid whose faces are regular hexagons and squares is a zonohedron. Thus Hadwiger's result can be formulated as follows: *A convex polyhedron A is T-equidecomposable with a cube if and only if A is a zonohedron* (i.e., the Minkowski sum of several line segments). It follows that two zonohedra of equal volume are T-equidecomposable. In particular, two congruent zonohedra are T-equidecomposable, no matter how they have been rotated with respect to each other.

§20. The Dehn–Hadwiger invariants and Jessen's theorem

A multidimensional generalization of the Dehn invariants has been proposed by Hadwiger [27]. To explain Hadwiger's ideas, we first agree to let $\varphi_0 (B)$ denote the p-dimensional volume of the p-dimensional polyhedron B (for arbitrary $p = 1, 2, \ldots$). Then the Dehn invariant of a polyhedron $A \subset R^3$ can be written in the form

$$f(A) = \sum l_i f(\alpha_i) = \sum \varphi_0 (A_i) f(\alpha_i), \qquad (112)$$

where $A_1, \ldots, A_k$ are the edges (i.e., one-dimensional faces) of the polyhedron A, with lengths $l_1 = \varphi_0 (A_1), \ldots, l_k = \varphi_0 (A_k)$, respectively, $\alpha_1, \ldots, \alpha_k$ are the corresponding dihedral angles, and $f(x)$ is a Cauchy–Hamel function satisfying the condition $f(\pi) = 0$.

The proof of the *additivity* of the function $f(A)$ (Lemma 11) was based on the following considerations: Let $A = P_1 + \ldots + P_q$. The edges (one-dimensional faces) of the polyhedra A, $P_1, \ldots, P_q$ can be divided into links in such a way that the polyhedra A, P_1, $\ldots, P_q$ adjoin one another along whole links. Moreover, φ_0 is an additive function, i.e., the length $l_i = \varphi_0 (A_i)$ of the edge A_i equals the *sum* of the lengths of the links making up A_i. Finally, the sum $\gamma_1 + \ldots + \gamma_s$ of the dihedral angles at a link m, taken over all the polyhedra P_i adjoining m, equals 2π or π if m does not lie on any edge of the polyhedron A, while the sum equals α_i or $\alpha_i - \pi$ if m lies on the edge A_i of the polyhedron A.

It is easy to see that all these considerations are immediately applicable to n-dimensional polyhedra for every $n \geqslant 3$. In fact, let $A \subset R^n$ be an n-dimensional polyhedron, and let $A_1, \ldots, A_k$ be all its $(n - 2)$-dimensional faces. If Π is a two-dimensional plane in R^n orthogonal to the face A_i and passing through an interior point q of A_i, then $A \cap \Pi$ is a polygon in the plane Π, with q as one of its vertices. The angle of the polygon $A \cap \Pi$ at the vertex q is taken to be the *dihedral angle* of the polyhedron A at its $(n - 2)$-dimensional face A_i.

Now let $\alpha_1, \ldots, \alpha_k$ denote the dihedral angles of the polyhedron A at its $(n - 2)$-dimensional faces $A_1, \ldots, A_k$. Moreover, let f be a Cauchy–Hamel function satisfying the condition $f(\pi) = 0$, and let φ be an additive D-invariant defined for $(n - 2)$-dimensional polyhedra. Then the function

$$f(A) = \sum_{i=1}^{k} \varphi(A_i) f(\alpha_i), \tag{113}$$

defined on the set of all n-dimensional polyhedra, is obviously D-invariant.

If $A = P_1 + \ldots + P_r$, then, in general, the polyhedra $A, P_1, \ldots, P_r$ do not adjoin one another along *whole* $(n - 2)$-dimensional faces. However, by taking all possible $(n - 2)$-dimensional polyhedra which are intersections of $(n - 2)$-dimensional faces of the polyhedra $A, P_1, \ldots, P_r$, we get a finite number of smaller polyhedra (*cells*), and the polyhedra $A, P_1, \ldots, P_r$ then adjoin one another along whole cells. It can now be shown (cf. the proof of Lemma 11) that the function (113) is additive. Thus we have the following

Theorem 28. *Let f be a Cauchy–Hamel function satisfying the condition $f(\pi) = 0$, and let φ be an additive D-invariant defined for $(n - 2)$-dimensional polyhedra. Then the function* (113) *(summed over all $(n - 2)$-dimensional faces of the polyhedron A) is an additive D-invariant.*

In particular, if the function $\varphi = \varphi_0$ is $(n - 2)$-dimensional volume, we get the additive D-invariant

$$\varphi_1(A) = \sum \varphi_0(A_i) f(\alpha_i). \tag{114}$$

For $n = 3$ this formula gives the Dehn invariant of a polyhedron $A \subset R^3$.

Formula (14) also defines the function φ_1 on the set of all *k-dimensional* polyhedra for $3 \leqslant k \leqslant n$ (in this case, the summation is over the $(k - 2)$-dimensional faces, and φ_0 denotes $(k - 2)$-dimensional volume). In particular, we can consider φ_1 for $(n - 2)$-dimensional polyhedra, i.e.,

$$\varphi_1(A) = \sum \varphi_0(A_i) f_0(\alpha_i), \qquad (115)$$

where f_0 is a Cauchy–Hamel function satisfying the condition $f_0(\pi) = 0$ (the summation is over the $(n - 4)$-dimensional faces of the $(n - 2)$-dimensional polyhedron A).

If $n \geqslant 5$, we can now choose φ_1 for the function φ in (113), thereby obtaining the additive D-invariant

$$\varphi_2(A) = \sum \varphi_1(A_i) f_1(\alpha_i),$$

where φ_1 is given by formula (115), and f_1 is a Cauchy–Hamel function satisfying the condition $f_1(\pi) = 0$ (and, in general, different from f_0). The invariant φ_2 can in turn be considered on k-dimensional polyhedra $(5 \leqslant k \leqslant n)$, and then, choosing φ_2 for the invariant φ in (113), we get the additive D-invariant

$$\varphi_3(A) = \sum \varphi_2(A_i) f_2(\alpha_i),$$

where f_2 is a new Cauchy–Hamel function, and so on. In general, the formula

$$\varphi_j(A) = \sum \varphi_{j-1}(A_i) f_{j-1}(\alpha_i),$$

defines the invariants $\varphi_j(A)$, $j = 1, \ldots, r$ in R^n, where r is the largest integer satisfying the condition $2r + 1 \leqslant n$.

The functions $\varphi_1, \ldots, \varphi_r$, introduced in Hadwiger's paper [27], will be called the *Dehn–Hadwiger invariants*. Since $\varphi_1, \ldots, \varphi_r$ are additive D-invariants, it follows from Theorem 15 that *a necessary condition for equidecomposability of two n-dimensional polyhedra M, $N \subset R^n$ is that the equalities*

$$\varphi_j(M) = \varphi_j(N), \qquad j = 0, 1, \ldots, r, \qquad (116)$$

hold (the first of these expresses the requirement of *equality of volume*).

Hadwiger [27] used the invariant φ_1 to generalize Theorem 20: *For* $n \geqslant 3$ *the regular* n-*dimensional simplex is not equidecomposable with a cube.* In fact, the dihedral angle at the $(n - 2)$-dimensional face of a regular n-dimensional simplex A equals arc cos $1/n$, and by Lemma 9, there exists a Cauchy–Hamel function f such that $f(\pi) = 0$, f(arc cos $1/n) \neq 0$. Using this function to define the invariant φ_1 (see (114)), we find that $\varphi_1(A) \neq 0$, while at the same time $\varphi_1(B) = 0$ for a cube B. Therefore A and B are nonequidecomposable.

In R^3 and R^4 there is only one Dehn–Hadwiger invariant φ_1 other than the volume φ_0. As shown by the Dehn–Sydler theorem, the necessary condition (116) for equidecomposability is also sufficient in three-dimensional space. Jessen [37] has proved that the same result is valid in four-dimensional space:

Theorem 29. *A necessary and sufficient condition for the equidecomposability of two four-dimensional polyhedra M and N of equal volume is that their Dehn–Hadwiger invariants* $\varphi_1(M)$ *and* $\varphi_1(N)$ *(see (114)) be equal for every Cauchy–Hamel function f satisfying the condition* $f(\pi) = 0$.

The proof uses the following three lemmas (see [37]):

Lemma 32. *For every 2-prism* $A \times B \subset R^4$ *and every* $\lambda > 0$, *there exist polyhedra P, Q* $\in Z_3$ *such that*

$$\lambda A \times B + P \underset{T}{\sim} A \times \lambda B + Q.$$

PROOF. We have (cf. (97))

$$[e_1] \times \lambda [e_2, e_3, e_4] + [e_1, e_2] \times \lambda [e_3, e_4] +$$
$$+ [e_1, e_2, e_3] \times \lambda [e_4] \underset{T}{\sim} \lambda [e_1] \times [e_2, e_3, e_4] +$$
$$+ \lambda [e_1, e_2] \times [e_3, e_4] + \lambda [e_1, e_2, e_3] \times [e_4]. \quad (117)$$

Replacing e_1 by $-e_1$ and interchanging the left- and right-hand sides, we obtain

$$\lambda\,[e_1] \times [e_2,\,e_3,\,e_4] + \lambda\,[-e_1,\,e_2] \times [e_3,\,e_4] +$$
$$+ \lambda\,[-e_1,\,e_2,\,e_3] \times [e_4] \underset{T}{\sim} [e_1] \times \lambda\,[e_2,\,e_3,\,e_4] +$$
$$+ [-e_1,\,e_2] \times \lambda\,[e_3,\,e_4] + [-e_1,\,e_2,\,e_3] \times \lambda\,[e_4]$$

(bearing in mind that $[-e_1] \underset{T}{\cong} [e_1]$). Adding these two formulas and using Theorem 22 to drop identical terms in the left and right-hand sides, we find that

$$[e_1,\,e_2] \times \lambda\,[e_3,\,e_4] + \lambda\,[-e_1,\,e_2] \times [e_3,\,e_4] +$$
$$+ [e_1,\,e_2,\,e_3] \times \lambda\,[e_4] + \lambda\,[-e_1,\,e_2,\,e_3] \times [e_4] \underset{T}{\sim}$$
$$\underset{T}{\sim} \lambda\,[e_1,\,e_2] \times [e_3,\,e_4] + [-e_1,\,e_2] \times \lambda\,[e_3,\,e_4] +$$
$$+ \lambda\,[e_1,\,e_2,\,e_3] \times [e_4] + [-e_1,\,e_2,\,e_3] \times \lambda\,[e_4].$$

Replacing e_4 by $-e_4$ here and interchanging the left- and right-hand sides, we get another formula. Adding this formula to the preceding one and dropping identical terms in the left- and right-hand sides, we obtain

$$[e_1,\,e_2] \times \lambda\,[e_3,\,e_4] + [-e_1,\,e_2] \times \lambda\,[e_3,\,-e_4] +$$
$$+ \lambda\,[-e_1,\,e_2] \times [e_3,\,e_4] + \lambda\,[e_1,\,e_2] \times [e_3,\,-e_4] \underset{T}{\sim}$$
$$\underset{T}{\sim} [-e_1,\,e_2] \times \lambda\,[e_3,\,e_4] + [e_1,\,e_2] \times \lambda\,[e_3,\,-e_4] +$$
$$+ \lambda\,[e_1,\,e_2] \times [e_3,\,e_4] + \lambda\,[-e_1,\,e_2] \times [e_3,\,-e_4]. \qquad (118)$$

We now note that

$$[-e_1,\,e_2] + [e_1 - e_2,\,2e_2] \underset{T}{\sim} [e_1,\,e_2] + [e_1] \times [e_2]$$

(Fig. 108), and hence

$$[-e_1,\,e_2] \times \lambda\,[e_3,\,e_4] + [e_1 - e_2,\,2e_2] \times \lambda\,[e_3,\,e_4] \underset{T}{\sim}$$

$$\underset{T}{\sim} [e_1,\,e_2] \times \lambda\,[e_3,\,e_4] + [e_1] \times [e_2] \times \lambda\,[e_3,\,e_4].$$

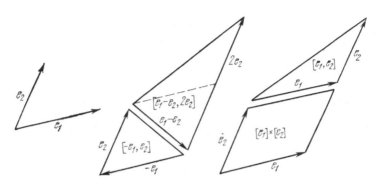

FIG. 108

Similarly,

$$[e_1, e_2] \times \lambda [e_3, -e_4] + [e_1] \times [e_2] \times \lambda [e_3, -e_4] \underset{T}{\sim}$$
$$\underset{T}{\sim} [-e_1, e_2] \times \lambda [e_3, -e_4] + [e_1 - e_2, \ 2e_2] \times \lambda [e_3, -e_4],$$
$$\lambda [e_1, e_2] \times [e_3, e_4] + \lambda [e_1] \times \lambda [e_2] \times [e_3, e_4] \underset{T}{\sim}$$
$$\underset{T}{\sim} \lambda [-e_1, c_2] \times [e_3, e_4] + \lambda [e_1 - e_2, 2e_2] \times [e_3, e_4],$$
$$\lambda [-e_1, e_2] \times [e_3, -e_4] + \lambda [e_1 - e_2, 2e_2] \times [e_3, -e_4] \underset{T}{\sim}$$
$$\underset{T}{\sim} \lambda [e_1, e_2] \times [e_3, -e_4] + \lambda [e_1] \times \lambda [e_2] \times [e_3, -e_4].$$

Adding these four formulas to (118) and simplifying, we find that

$$[e_1 - e_2, \ 2e_2] \times \lambda [e_3, e_4] + \lambda [e_1 - e_2, 2e_2] \times [e_3, -e_4] + P' \underset{T}{\sim}$$
$$\underset{T}{\sim} \lambda [e_1 - e_2, \ 2e_2] \times [e_3, e_4] +$$
$$+ [e_1 - e_2, 2e_2] \times \lambda [e_3, -e_4] + Q', \quad (119)$$

where P', $Q' \in Z_3$. Moreover, since

$$[e_3, -e_4] + [2e_3, e_4 - e_3] \underset{T}{\sim} [e_3, e_4] + [e_3] \times [e_4],$$

we have

$$[e_1 - e_2, 2e_2] \times \lambda [e_3, -e_4] + [e_1 - e_2, 2e_2] \times \lambda [2e_3, e_4 - e_3] \underset{T}{\sim}$$

$$\underset{T}{\sim} [e_1 - e_2, 2e_2] \times \lambda [e_3, e_4] + [e_1 - e_2, 2e_2] \times \lambda [e_3] \times \lambda [e_4],$$

$$\lambda [e_1 - e_2, 2e_2] \times [e_3, e_4] + \lambda [e_1 - e_2, 2e_2] \times [e_3] \times [e_4] \underset{T}{\sim}$$

$$\underset{T}{\sim} \lambda [e_1 - e_2, 2e_2] \times [e_3, -e_4] + \lambda [e_1 - e_2, 2e_2] \times$$

$$\times [2e_3, e_4 - e_3].$$

Adding these formulas to (119) and simplifying, we get

$$[e_1 - e_2, 2e_2] \times \lambda [2e_3, e_4 - e_3] + P'' \underset{T}{\sim}$$

$$\underset{T}{\sim} \lambda [e_1 - e_2, 2e_2] \times [2e_3, e_4 - e_3] + Q'',$$

$$P'', Q'' \in Z_3.$$

Here e_1, e_2, e_3, e_4 are arbitrary linearly independent vectors. Setting $e_1 - e_2 = a$, $2e_2 = b$, $2e_3 = c$, $e_4 - e_3 = d$, we find that

$$[a, b] \times \lambda [c, d] + P'' \underset{T}{\sim} \lambda [a, b] \times [c, d] + Q''$$

(for arbitrary linearly independent a, b, c, d). In particular,

$$\lambda [e_1, e_2] \times [e_3, e_4] + P_1 \underset{T}{\sim} [e_1, e_2] \times \lambda [e_3, e_4] + Q_1;$$

$$P_1, Q_1 \in Z_3. \qquad (120)$$

Addition of this formula to (117) gives

$$[e_1] \times \lambda [e_2, e_3, e_4] + [e_1, e_2, e_3] \times \lambda [e_4] + P_1 \underset{T}{\sim}$$

$$\underset{T}{\sim} \lambda [e_1] \times [e_2, e_3, e_4] + \lambda [e_1, e_2, e_3] \times [e_4] + Q_1.$$

Replacing e_1 by $-e_1$ and interchanging the left- and right-hand sides, we find that

$$\lambda\,[e_1] \times [e_2,\, e_3,\, e_4] + \lambda\,[-e_1,\, e_2,\, e_3] \times [e_4] + P_1' \underset{T}{\sim}$$

$$\underset{T}{\sim} [e_1] \times \lambda\,[e_2,\, e_3,\, e_4] + [-e_1,\, e_2,\, e_3] \times \lambda\,[e_4] + Q_1'.$$

Adding this formula to the preceding one, we get

$$[e_1,\, e_2,\, e_3] \times \lambda\,[e_4] + \lambda\,[-e_1,\, e_2,\, e_3] \times [e_4] + P_2 \underset{T}{\sim}$$

$$\underset{T}{\sim} \lambda\,[e_1,\, e_2,\, e_3] \times [e_4] + [-e_1,\, e_2,\, e_3] \times \lambda\,[e_4] + Q_2 \quad (121)$$

(where $P_2,\, Q_2 \in Z_3$).

Next we introduce the notation $\{b_1,\, b_2,\, b_3\}$ for the tetrahedron constructed on the vectors $b_1,\, b_2,\, b_3$, i.e., the tetrahedron with vertices $b_0,\, b_0 + b_1,\, b_0 + b_2,\, b_0 + b_3$ (where b_0 is chosen arbitrarily). It is not hard to verify that

$$2 \cdot \{e_1,\, e_2,\, e_3\} + [-e_1,\, b_2,\, b_3] \underset{T}{\sim} \{2e_1,\, e_2,\, e_3\} + [e_1,\, b_2,\, b_3],$$

where $b_2 = e_2 - e_1$, $b_3 = e_3 - e_2$. From this we obtain

$$2 \cdot \{e_1,\, e_2,\, e_3\} \times [e_4] + [-e_1,\, b_2,\, b_3] \times [e_4] \underset{T}{\sim}$$
$$\underset{T}{\sim} \{2e_1,\, e_2,\, e_3\} \times [e_4] + [e_1,\, b_2,\, b_3] \times [e_4] \quad (122)$$

(for arbitrary linearly independent $e_1,\, e_2,\, e_3,\, e_4$). Replacing e_4 by λe_4 in this formula and interchanging the left- and right-hand sides, we get

$$\{2e_1,\, e_2,\, e_3\} \times \lambda[e_4] + [e_1,\, b_2,\, b_3] \times \lambda\,[e_4] \underset{T}{\sim}$$

$$\underset{T}{\sim} 2 \cdot \{e_1,\, e_2,\, e_3\} \times \lambda\,[e_4] + [-e_1,\, b_2,\, b_3] \times \lambda\,[e_4].$$

Moreover, replacing $e_1,\, e_2,\, e_3$ in (122) by $\lambda e_1,\, \lambda e_2,\, \lambda e_3$, we have

$$2 \cdot \lambda\,\{e_1,\, e_2,\, e_3\} \times [e_4] + \lambda\,[-e_1,\, b_2,\, b_3] \times [e_4] \underset{T}{\sim}$$

$$\underset{T}{\sim} \lambda\,\{2e_1,\, e_2,\, e_3\} \times [e_4] + \lambda\,[e_1,\, b_2,\, b_3] \times [e_4].$$

Adding the last two formulas and taking account of (121), we find that

$$\{2e_1, e_2, e_3\} \times \lambda [e_4] + 2 \cdot \lambda \{e_1, e_2, e_3\} \times [e_4] + Q_2 \underset{T}{\sim}$$

$$\underset{T}{\sim} 2 \cdot \{e_1, e_2, e_3\} \times \lambda [e_4] + \lambda \{2e_1, e_2, e_3\} \times [e_4] + P_2, \quad (123)$$

where $P_2, Q_2 \in Z_3$. Similarly,

$$\{2e_1, 2e_2, e_3\} \times \lambda [e_4] + 2 \cdot \lambda \{2e_1, e_2, e_3\} \times [e_4] + Q_2' \underset{T}{\sim}$$

$$\underset{T}{\sim} 2 \cdot \{2e_1, e_2, e_3\} \times \lambda [e_4] + \lambda \{2e_1, 2e_2, e_3\} \times [e_4] + P_2', \quad (124)$$

$$\{2e_1, 2e_2, 2e_3\} \times \lambda [e_4] + 2 \cdot \lambda \{2e_1, 2e_2, e_3\} \times [e_4] + Q_2'' \underset{T}{\sim}$$

$$\underset{T}{\sim} 2 \cdot \{2e_1, 2e_2, e_3\} \times \lambda [e_4] + \lambda \{2e_1, 2e_2, 2e_3\} \times [e_4] + P_2''.$$
$$(125)$$

Adding formula (125), two copies of formula (124) and four copies of formula (123), we now obtain

$$8 \cdot \lambda \{e_1, e_2, e_3\} \times [e_4] + \{2e_1, 2e_2, 2e_3\} \times \lambda [e_4] + Q^* \underset{T}{\sim}$$

$$\underset{T}{\sim} 8 \cdot \{e_1, e_2, e_3\} \times \lambda [e_4] + \lambda \{2e_1, 2e_2, 2e_3\} \times [e_4] + P^*;$$
$$P^*, Q^* \in Z_3. \quad (126)$$

Furthermore, taking account of the formula $8 \cdot M + \tilde{P} \underset{T}{\sim} 2M + 6M + \tilde{Q}$, where $\tilde{P}, \tilde{Q} \in Z_2$, implied by Lemma 16, we can write

$$8 \cdot \{e_1, e_2, e_3\} \times \lambda [e_4] + Q_1^* \underset{T}{\sim}$$

$$\underset{T}{\sim} \{2e_1, 2e_2, 2e_3\} \times \lambda [e_4] + \{6e_1, 6e_2, 6e_3\} \times \lambda [e_4] + P_1^*,$$

$$\lambda \{2e_1, 2e_2, 2e_3\} \times [e_4] + \lambda \{6e_1, 6e_2, 6e_3\} \times [e_4] + Q_2^* \underset{T}{\sim}$$

$$\underset{T}{\sim} 8 \cdot \lambda \{e_1, e_2, e_3\} \times [e_4] + P_2^*,$$

where P_i^*, $Q_i^* \in Z_3$. Adding the last two formulas and formula (126), we finally obtain

$$\lambda \{6e_1, 6e_2, 6e_3\} \times [e_4] + Q^{**} \underset{T}{\sim}$$

$$\underset{T}{\sim} \{6e_1, 6e_2, 6e_3\} \times \lambda [e_4] + P^{**},$$

$$P^{**}, Q^{**} \in Z_3. \quad (127)$$

Formulas (120) and (127) show that the assertion of Lemma 32 is valid in the case where A and B are *simplexes*. From this we conclude, by making a decomposition into simplexes, that the assertion is valid for an arbitrary 2-prism $A \times B \subset R^4$.

Lemma 33. *Every polyhedron* $A \subset R^4$ *is equidecomposable with a polyhedron of the form* $I \times W$, *where* I *is a segment of length* 1 *orthogonal to the three-dimensional polyhedron* W.

PROOF. First we consider an arbitrary *orthogonal simplex in* R^4, i.e., a simplex $[e_1, e_2, e_3, e_4]$ where e_1, e_2, e_3, e_4 are pairwise orthogonal vectors (of arbitrary length). Every point x of this simplex has a unique representation of the form

$$x = b_0 + x_1 e_1 + x_2 e_2 + x_3 e_3 + x_4 e_4 \quad (128)$$

(see (41)), where the coordinates x_1, x_2, x_3, x_4 satisfy the conditions

$$1 \geqslant x_1 \geqslant x_2 \geqslant x_3 \geqslant x_4 \geqslant 0. \quad (129)$$

Let M denote the set of points $x \in [e_1, e_2, e_3, e_4]$ satisfying the extra condition $x_2 + x_3 \leqslant 1$ and N the set of points $x \in [e_1, e_2, e_3, e_4]$ for which $x_2 + x_3 \geqslant 1$. It is clear that

$$[e_1, e_2, e_3, e_4] = M + N.$$

Moreover, let M_1 denote the set of points $x \in M$ satisfying the condition $x_2 \geqslant \frac{1}{2}$ and M_2 the set of points $x \in M$ for which $x_2 \geqslant \frac{1}{2}$. Then $M = M_1 + M_2$. Let g be the mapping carrying the point (128) into the point

g be the mapping carrying the point (128) into the point

$$b_0 + (1 - x_1)e_1 + (1 - x_2)e_2 + x_3e_3 + x_4e_4. \tag{130}$$

Because of the orthogonality of the vectors e_1, e_2, e_3, e_4, the mapping g is a *motion*. The polyhedron $g(M_2)$ consists of the points (130) satisfying the conditions (129) and the inequalities $x_2 + x_3 \leqslant 1$, $x_2 \geqslant {}^1\!/_2$; in other words, $g(M_2)$ consists of the points (128) satisfying the conditions

$$0 \leqslant x_1 \leqslant x_2, \quad 0 \leqslant x_4 \leqslant x_3 \leqslant x_2 \leqslant {}^1\!/_2.$$

From this it is clear that $M_1 + g(M_2)$ consists of the points (128) satisfying the conditions

$$0 \leqslant x_1 \leqslant 1, \quad 0 \leqslant x_4 \leqslant x_3 \leqslant x_2 \leqslant {}^1\!/_2,$$

i.e., $M_1 + g(M_2) = [e_1] \times {}^1\!/_2 [e_2, e_3, e_4]$.

Similarly, dividing N into two parts N_1, N_2 by the conditions $x_3 \leqslant {}^1\!/_2$, $x_3 \geqslant {}^1\!/_2$, and applying the motion h carrying the point (128) into the point

$$b_0 + x_1e_1 + x_2e_2 + (1 - x_3)e_3 + (1 - x_4)e_4,$$

we find that $h(N_1) + N_2 = {}^1\!/_2 [e_1, e_2, e_3] \times [e_4]$. Thus

$$[e_1, e_2, e_3, e_4] \sim [e_1] \times {}^1\!/_2 [e_2, e_3, e_4] + {}^1\!/_2 [e_1, e_2, e_3] \times [e_4]. \tag{131}$$

Now let λ be the length of the vector e_1, so that $e_1 = \lambda a_1$, where a_1 is a vector of length 1, orthogonal to the three-dimensional space $L \subset R^4$ spanned by the vectors e_2, e_3, e_4. By Lemma 32,

$$[e_1] \times \frac{1}{2} [e_2, e_3, e_4] + P \underset{T}{\sim} [a_1] \times \frac{\lambda}{2} [e_2, e_3, e_4] + Q,$$

$$P, Q \in Z_3. \tag{132}$$

Moreover, every 3-prism in R^4 is of the form $A \times B \times C$, where A, B are line segments and C is a polygon; since the polygon C is equidecomposable with a parallelogram (Theorem 10), the 3-prism $A \times B \times C$ is equidecomposable with some four-dimensional *parallelepiped*. Hence, for the polyhedra P, Q (see (132)) we have

$$P \sim P', \quad Q \sim Q',$$

where P', $Q' \in Z_4$. By Lemma 18, we can assume that

$$P' = [a_1] \times U, \quad Q' = [a_1] \times V,$$

where U, $V \subset L$ are parallelepipeds. Thus (132) takes the form

$$[e_1] \times \tfrac{1}{2}[e_2, e_3, e_4] + [a_1] \times U \sim [a_1] \times \left(\frac{\lambda}{2}[e_2, e_3, e_4] + V \right).$$

But $v(U) < v\left(\dfrac{\lambda}{2}[e_2, e_3, e_4] + V \right)$, since the polyhedra in the left- and right-hand sides have the same volume, and therefore

$$\frac{\lambda}{2}[e_2, e_3, e_4] + V = U' + W',$$

where $U' \underset{T}{\sim} U$ (Lemma 12). Thus

$$[e_1] \times \tfrac{1}{2}[e_2, e_3, e_4] + [a_1] \times U \sim [a_1] \times U' + [a_1] \times W',$$

i.e., the polyhedra $[e_1] \times \tfrac{1}{2}[e_2, e_3, e_4]$ and $[a_1] \times W'$ are equi-complementable, and hence equidecomposable (by Theorem 22):

$$[e_1] \times \tfrac{1}{2}[e_2, e_3, e_4] \sim I \times W'.$$

Here $W' \subset L$ is a three-dimensional polyhedron, and $I = [a_1]$ is a segment of length 1 orthogonal to the subspace L. Similarly,

$$\tfrac{1}{2}[e_1, e_2, e_3] \times [e_4] \sim I \times W'',$$

where we can assume (applying a motion, if necessary) that W' and W'' lie in one and the same subspace L and have no common interior points. Thus, by (131),

$$[e_1, e_2, e_3, e_4] \sim I \times (W' + W'');$$

i.e., Lemma 33 is valid in the case where A is an orthogonal simplex.

Moreover, if A is an arbitrary simplex, then there exist orthogonal simplexes A_i', A_j'', such that

$$A + A_1' + \ldots + A_k' \underset{T}{\sim} A_1'' + \ldots + A_l'' \qquad (133)$$

(cf. the proof of Lemma 29). By what was just proved, it follows from (133) that the assertion of Lemma 33 is valid for an *arbitrary* simplex, and hence for an arbitrary polyhedron.

Lemma 34. *Let W be a three-dimensional polyhedron, lying in a three-dimensional subspace $L \subset R^4$, and let I be a segment of length 1 orthogonal to L. Then $\varphi_1 (I \times W) = \varphi_1 (W)$, where φ_1 is the Dehn–Hadwiger invariant* (see (114)).

PROOF. Let a and b denote the end points of the segment I. Then the two-dimensional faces of the polyhedron $I \times W$ are polygons of the form $a \times M$, $b \times M$, $I \times K$, where M is an arbitrary two-dimensional face of the polyhedron W, and K is an arbitrary edge of W. The dihedral angles at the faces $a \times M$ and $b \times M$ are right angles, and hence these faces need not be considered in calculating the invariant $\varphi_1 (I \times W)$. Moreover, the dihedral angle of the polyhedron $I \times W$ at the face $I \times K_i$ equals the dihedral angle α_i of the polyhedron W at the edge K_i. Therefore, denoting the length of the edge K_i by l_i, we have

$$\varphi_0 (I \times K_i) f (\alpha_i) = s (I \times K_i) f (\alpha_i) = (1 \cdot l_i) f (\alpha_i)$$
$$= l_i f (\alpha_i),$$

whence, summing over all the edges of the polyhedron W, we get the required formula $\varphi_1 (I \times W) = \varphi_1 (W)$.

Proof of Theorem 29. The necessity follows from the fact that φ_1

is an additive D-invariant (Theorem 28). To prove the sufficiency, let $A, B \subset R^4$ be four-dimensional polyhedra such that

$$\varphi_0 (A) = \varphi_0 (B), \quad \varphi_1 (A) = \varphi_1 (B)$$

(for an arbitrary choice of the Cauchy–Hamel function f satisfying the condition $f (\pi) = 0$). By Lemma 33,

$$A \sim I \times W_1, \quad B \sim I \times W_2,$$

where W_1, W_2 are three-dimensional polyhedra lying in a subspace $L \subset R^4$ and I is a segment of length 1 orthogonal to L. We have

$$v (W_1) = \varphi_0 (I \times W_1) = \varphi_0 (A) = \varphi_0 (B) = \varphi_0 (I \times W_2)$$
$$= v (W_2).$$

Moreover, according to Lemma 34,

$$\varphi_1 (W_1) = \varphi_1 (I \times W_1) = \varphi_1 (A) = \varphi_1 (B) = \varphi_1 (I \times W_2)$$
$$= \varphi_1 (W_2).$$

Thus W_1 and W_2 are polyhedra of equal volume, whose Dehn invariants are equal, and hence $W_1 \sim W_2$ (Theorem 25). But then $I \times W_1 \sim I \times W_2$, so that $A \sim B$.

§21. Minimality of the group of orientation-preserving motions

As we saw in §12, the concepts of D-equidecomposability and D_0-equidecomposability are equivalent. Here we prove [8] that D_0 is the *minimal* group with this property, i.e., the following theorem is valid (analogous to Theorem 17 for polygons):

Theorem 30. *Let G be a group of motions of the space R^3 such that any two equidecomposable polyhedra are G-equidecomposable. Then G coincides with one of the groups D, D_0.*

In other words, if G does not coincide with one of the groups D, D_0, we can then find two equidecomposable polyhedra which are not G-equidecomposable.

PROOF. Let $T^* \subset R^3$ denote the tetrahedron defined by the inequalities

$$x \geqslant 0,\ y \geqslant 0,\ z \geqslant 0;\ \sqrt{\frac{2-\sqrt{3}}{4}}\,(x+y) + \sqrt{\frac{\sqrt{3}}{2}}\,z \leqslant 1$$

in a system of rectangular coordinates x, y, z. Let o denote the coordinate origin and a, b, c the vertices of the tetrahedron T^* lying on the x, y, z axes. The unit vectors of the exterior normals to the faces of the tetrahedron T^* are of the form

$$a_1 = (-1, 0, 0);\quad a_2 = (0, -1, 0);\quad a_3 = (0, 0, -1);$$
$$a_4 = \left(\sqrt{\frac{2-\sqrt{3}}{4}},\ \sqrt{\frac{2-\sqrt{3}}{4}},\ \sqrt{\frac{\sqrt{3}}{2}}\right).$$

Therefore, letting $\gamma_1, \gamma_2, \gamma_3$ denote the dihedral angles of the tetrahedron T^* at the edges ab, ac, bc, we have

$$\cos \gamma_1 = -a_3 a_4 = \sqrt{\frac{\sqrt{3}}{2}},$$
$$\cos \gamma_2 = \cos \gamma_3 = -a_1 a_4 = \sqrt{\frac{2-\sqrt{3}}{4}}.$$

Since

$$\cos \frac{5\pi}{12} = \sqrt{\frac{1+\cos \frac{5\pi}{6}}{2}} = \sqrt{\frac{1-\frac{\sqrt{3}}{2}}{2}} = \sqrt{\frac{2-\sqrt{3}}{4}},$$

we have $\gamma_2 = \gamma_3 = 5\pi/12$, i.e., the dihedral angles at the edges ac and bc are rational multiples of the number π. The same is true of the dihedral angles at the edges oa, ob, oc (which are right angles).

We must still consider the dihedral angle at the edge *ab*. First we note that

$$\cos 2\gamma_1 = 2\cos^2 \gamma_1 - 1 = \sqrt{3} - 1.$$

The formula

$$\cos 2 (k + 1)\gamma_1 = 2 \cos 2\gamma_1 \cos 2k\gamma_1 - \cos 2 (k - 1)\gamma_1 \quad (134)$$

allows us to successively calculate $\cos 2k\gamma_1$ for $k = 1, 2, \ldots,$ starting from the values

$$\cos 2\gamma_1 = \sqrt{3} - 1; \quad \cos 4\gamma_1 = 2\cos^2 2\gamma_1 - 1 = -4\sqrt{3}+7.$$

In fact, we can set

$$\cos 2k\gamma_1 = (-1)^{k-1}p_h \sqrt{3} + (-1)^k q_h, \quad k = 1, 2, \ldots, \quad (135)$$

where, because of (134), p_h and q_h are given by the recursion relations

$$\begin{cases} p_{k+1} = 2p_k + 2q_k - p_{k-1}, & p_1 = 1, \ p_2 = 4, \\ q_{k+1} = 6p_k + 2q_k - q_{k-1}; & q_1 = 1, \ q_2 = 7. \end{cases}$$

Writing these relations in the form

$$p_{k+1} - p_k = (p_k - p_{k-1}) + 2q_k; \quad p_{k+1} = p_k + 2q_k + (p_k - p_{k-1});$$
$$q_{k+1} - q_k = (q_k - q_{k-1}) + 6p_k; \quad q_{k+1} = 6p_k + q_k + (q_k - q_{k-1}),$$

we find that if the inequalities

$$p_h - p_{h-1} > 0, \quad q_h - q_{h-1} > 0, \quad p_h > 0, \quad q_h > 0 \quad (136)$$

hold for some $k \geqslant 2$, then they also hold for $k + 1$. But the inequalities hold for $k = 2$, and hence they hold for all $k \geqslant 2$. It follows that p_k, q_k are positive integers, with $p_k \neq 0$ for $k = 1$, 2, ..., and hence, by (135), $\cos 2k\gamma_1$ is an *irrational* number. In particular, $\cos 2k\gamma_1 \neq 1$ for $k = 1$, 2, ..., i.e., $2k\gamma_1 \neq 2l\pi$ for positive integers k, l. In other words, the number γ_1/π is *irrational*. Therefore, setting

$$f(\pi) = f\left(\frac{\pi}{2}\right) = f\left(\frac{5\pi}{12}\right) = 0, \ f(\gamma_1) = 1, \qquad (137)$$

we get an *additive* function on the set $\{\pi, \ \pi/2, \ 5\pi/12, \ \gamma_1\}$, containing the number π and the dihedral angles of the tetrahedron T^*. The corresponding Dehn invariant $f(T^*)$ is nonzero, i.e.,

$$f(T^*) = lf(\gamma_1) = l \neq 0,$$

where l is the length of the edge ab.

Now let $p \subset R^3$ be an arbitrary line, and let P denote the set of all lines obtained from p by motions belonging to the group G. Moreover, let f be an additive function satisfying the condition $f(\pi) = 0$, and let M be an arbitrary polyhedron. We set

$$f_P(M) = l_1 f(\alpha_1) + \ldots + l_h f(\alpha_h),$$

where $l_1, \ldots, l_h$ are the lengths of the edges of the polyhedron M which lie on lines belonging to the set P, and $\alpha_1, \ldots, \alpha_h$ are the corresponding dihedral angles. By a verbatim repetition of the considerations given in the proof of Lemma 11 and Theorem 19, it can be proved that if $f_P(M) \neq f_P(N)$, then M and N are not G-equidecomposable.

Suppose now that there exists a line p_1 which cannot be obtained from p by any motion belonging to the group G. Let T be the tetrahedron congruent to T^* such that the edge congruent to ab lies on the line p, and let the tetrahedron T_1 bear the same relation to the line p_1. Then for the function f defined by the formulas (137) we have

$$f(T) = lf(\gamma_1) = l \neq 0, \quad f(T_1) = 0$$

(since $p_1 \not\in P$), so that the congruent (and hence equidecomposable) tetrahedra T and T_1 are not G-equidecomposable. But this contradicts the choice of the group G.

Thus, given any two lines p and p_1, there exists a motion $g \in G$ satisfying the condition $g(p) = p_1$, and we must still prove that a group G with this property contains D_0. To this end, let p_1, p_2, p_3 be three lines, no two of which are parallel, and let $g_1, g_2 \in G$ be motions such that $g_1(p_1) = p_2$, $g_2(p_2) = p_3$. The motion $g_3 = g_2 \circ g_1$ satisfies the condition $g_3(p_1) = p_3$. Since $g_3 = g_2 \circ g_1$, at least one of the motions g_1, g_2, g_3 is orientation-preserving. Hence there exists an orientation-preserving motion $g^* \in G$, which carries some line q into a line not parallel to q, so that g^* is not a translation.

It will be recalled that an orientation-preserving motion of the space R^3 other than a translation is a *twist*, i.e., the composition of a rotation through an angle α about some directed line l ($0 < \alpha < 2\pi$ and the rotation is counterclockwise looking along the direction of l) and a translation by a vector λa, where a is the unit vector of the directed line l; we will denote such a twist by (l, α, λ). Thus the motion $g^* \in G$ just found is of the form $(l^*, \alpha^*, \lambda^*)$.

Next let p be an arbitrary line, and let $g \in G$ be the motion carrying the line l^* into p, where we choose the direction of p in such a way that g carries the directed line l^* into the directed line p. It is easy to see that the motion $h = g \circ g^* \circ g^{-1} \in G$ is the twist (p, α^*, λ^*) with the same α^*, λ^*. The motion h^{-1} is the twist $(p', \alpha^*, \lambda^*)$, where p' is the directed line obtained from p by reversing its direction. Thus for *every* directed line p, there is a twist $g_p = (p, \alpha^*, \lambda^*)$ in the group G.

We now let p be a fixed directed line and p' the oppositely directed line, while q is a *variable* directed line. If q coincides with p, then the motion $g_q \circ g_p$ is the twist $(p, 2\alpha^*, 2\lambda^*)$ with angle of rotation $2\alpha^*$, while if q coincides with p', the motion $g_q \circ g_p$ is the identity transformation. Examining the motion $g_q \circ g_p$ as the directed line q is continuously varied from the position p to the position p', we find, by considerations of continuity, that the group

G contains a twist with angle β for every β between 0 and $2\alpha^*$. But this in turn implies that the group G contains a twist with angle β for every β between 0 and 2π. Moreover, as the above argument shows, for every directed line p and every β, $0 < \beta < 2\pi$, the group G contains a twist (p, β, λ) for some λ.

To prove the relation $G \supset D_0$, we must still show that G contains all translations, this will imply that along with the twist (p, β, λ), for *some* λ, the group G contains the twists (p, β, λ) for *arbitrary* λ. Let p_1 and p_2 be parallel lines which do not coincide. By what has already been proved, the group G contains twists $g_1 = (p_1, \pi, \lambda_1)$ and $g_2 = (p_2, \pi, \lambda_2)$ for some λ_1, λ_2. The composition $t = g_2 \circ g_1$ is then a translation by a nonzero vector. Thus G contains a translation t by a vector e of length $l \neq 0$. Let e' be any vector of the same length l, and choose a twist $g \in G$ carrying the vector e into the vector e'. Then the motion $g \circ t \circ g^{-1} \in G$ is a translation by the vector e'. Thus G contains the translation by any vector of length l. Taking a composition of identical translations, we find that G contains the translation by any vector of length kl, where k is a positive integer. But then *every* translation belongs to the group G (to see this, we need only consider an isosceles triangle whose base is an arbitrarily chosen segment and whose lateral sides equal kl).

§22. The algebra of polyhedra

So far the notation $M = A_1 + \ldots + A_k$ has always meant that the polyhedra $A_1, \ldots, A_k$ have pairwise disjoint interiors, and that, under these conditions, M is the union of the polyhedra $A_1, \ldots, A_k$. In keeping with this, if it is required, for example, to add the formulas $A \sim B$ and $C \sim D$, we must first make sure that the polyhedra in each side of the formulas have no interior points in common, i.e., instead of writing $A + C \sim B + D$, we must consider the formula $A + C' \sim B + D'$, where the polyhedra $C' \cong C$, $D' \cong D$ are chosen in such a way that the sums $A + C'$ and $B + D'$ are defined. However, we might agree not to distinguish between congruent polyhedra, i.e., to regard the terms in each sum as being defined only to within congruence, which would allow us to write the sum $A + C$ (or $B + D$) without worrying about whether or not the terms have common interior points. Let us also interpret the formula $A \sim B - C$ as meaning $A + C \sim B$ (this interpretation of the difference has already been used in §17). Finally, agreeing to regard every 2-prism as equivalent to zero, we find, by Lemma 15, that the polyhedra $(\lambda + \mu)M$ and $\lambda M + \mu M$ are equivalent for arbitrary λ, μ; in fact, it is just such an equivalence relation that is the basis for the considerations of §17. The formalization of this equivalence relation leads to the concept of the *algebra of polyhedra,* introduced by Hadwiger [24], which is the topic to which the present section is devoted.

Let P^n be the vector space whose elements are the formal linear combinations $\lambda_1 M_1 + \ldots + \lambda_k M_k$, where k is any positive integer, $M_1, \ldots, M_k$ are arbitrary n-dimensional polyhedra in R^n, and $\lambda_1, \ldots, \lambda_k$ are arbitrary real numbers. Let G be a fixed group of motions of the space R^n, containing the group T of all translations. We now list four types of formal linear combinations (elements of P^n) which will be regarded as "*nonessential*": 1) elements of the type $M - N = 1 M + (-1)N$, where $M \underset{G}{\cong} N$, 2) |elements of the type $A - M_1 - \ldots - M_k$, where $A = M_1 \cup \ldots \cup M_k$ and $M_1, \ldots, M_k$ have pairwise disjoint interiors; 3) elements of the type $M - \lambda N$, where the polyhedron M is obtained from the polyhedron N by a homothetic transformation with ratio $\lambda > 0$; 4) all 2-prisms. Let $P_0^n (G)$ denote the subspace of P^n generated by all nonessential elements. Then the factor space $\Sigma^n(G) = P^n/P_0^n (G)$ will be called the n-dimensional *G-algebra of polyhedra.* Let $\varkappa$ denote the natural homomorphism $P^n \to P^n/P_0^n (G)$.

It is not hard to show, by using Lemma 15 and Theorem 22, that *given any two polyhedra A, $B \subset R^n$, the formula $\varkappa (A - B) = 0$* (*i.e., the relation $A - B \in P_0^n (G)$) holds if and only if A and B are G-equidecomposable modulo Z_2, i.e., if and only if there exist polyhedra U, $V \in Z_2$ such that $A + U$ and $B + V$ are G-equidecomposable.* It is just this fact that constitutes the geometric meaning of the algebra of polyhedra. We note that when $n = 3$ equidecomposability modulo Z_2 for polyhedra of equal volume *coincides* with ordinary equidecomposability (this follows from

Lemma 19 and Theorem 22). In other words, taking G to be the group of all motions of the space R^3, we find that two polyhedra A, $B \subset R^3$ are equidecomposable if and only if $v(A) = v(B)$ and $\varkappa(A - B) = 0$. This was actually the algebraic basis for the considerations of §17. We also note if G is taken to be the group D of all motions of the space R^4, then $P_0^4(G) = P^4$ (Lemma 33), i.e., the factor space $\Sigma^4(D) = P^4/P_0^4(D)$ is trivial. This served as the algebraic basis for Jessen's reasoning (Theorem 29). Moreover, as shown by Hadwiger [24], if n is *even* and if the group G of motions of the space R^n contains all translations and all central inversions, then $P^n = P_0^n(G)$, i.e., the space $\Sigma^n(G)$ is trivial. We also note that for $n - 3$ (and other odd values of n) the space $\Sigma^n(D) - P^n/P_0^n(D)$ is *infinite-dimensional* (more exactly, is of dimension $\aleph$). For $n = 3$ this follows from the fact (already established by Dehn [13]) that among three-dimensional polyhedra of equal volume there is a noncountable set of polyhedra which are pairwise nonequidecomposable.

In the three-dimensional case *the Dehn invariants are homomorphisms of the algebra of polyhedra* (i.e., linear functionals on the vector space $\Sigma^3(D) = P^3/P_0^3(D)$). In fact, if $f(A)$ is a Dehn invariant (defined by some Cauchy–Hamel function satisfying the condition $f(\pi) = 0$), we can write

$$f(\lambda_1 M_1 + \ldots + \lambda_k M_k) = \lambda_1 f(M_1) + \ldots + \lambda_k f(M_k)$$

for an arbitrary formal linear combination $\lambda_1 M_1 + \ldots + \lambda_k M_k \in P^3$. It can be verified immediately that f vanishes on the subspace $P_0^3(D)$, i.e., on every nonessential element (in particular, f vanishes on prisms, as we saw in §13). But this means that the Dehn invariant f can be regarded as a linear functional on the factor space $P^3/P_0^3(D)$. We note that this is no longer the case for $n > 3$, since the Dehn–Sydler invariants do not vanish on 2-prisms (cf. the proof of Theorem 29).

An interesting treatment of the Dehn invariants (in three-dimensional space) has been given by Jessen in [36]. Namely, let R_π denote the additive group of real numbers, reduced modulo π, and consider the tensor product $R \otimes R_\pi$. Moreover, let

$$\Delta \left(A \right) = l_1 \otimes \alpha_1 + \ldots + l_k \otimes \alpha_k,$$

where $l_1, \ldots, l_k$ are the lengths of the edges of the polyhedron $A \subset R^3$, and $\alpha_1, \ldots, \alpha_k$ are the corresponding dihedral angles. The function Δ can be extended, by additivity, onto the vector space P^3. It can easily be verified that the resulting homomorphism Δ: $P^3 \to R \otimes R_\pi$ vanishes on the subspace $P_0^3 (D)$, and hence defines a homomorphism $\delta = \Delta \circ \varkappa^{-1}$ on the factor space $\Sigma^3 (D) = P^3/P_0^3(D)$. The homomorphism Δ (or δ) actually incorporates all the Dehn invariants. More exactly, let $\varphi \colon R \otimes R_\pi \to R$ be an additive homomorphism which is linear in the first factor (i.e., which satisfies the condition $\varphi \left(\lambda \otimes \alpha \right) = \lambda \varphi \left(1 \otimes \alpha \right)$). Then $\varphi \circ \Delta \colon P^3 \to R$ is a Dehn invariant, and *every* Dehn invariant can be obtained in this way. With this approach the Dehn–Sydler theorem can be formulated as follows: *A necessary and sufficient condition for equidecomposability of two polyhedra $A, B \subset R^3$ is that* $v \left(A \right) = v \left(B \right)$, *and* $\Delta \left(A \right) = \Delta \left(B \right)$.

In conclusion, we mention one more problem connected with the algebra of polyhedra. The way of constructing the vector space $\Sigma^n \left(G \right) = P^n/P_0^n \left(G \right)$ immediately implies (with the help of Zorn's lemma) the existence in R^n of a set $B = \{N_\alpha\}$ of polyhedra, which form a *basis* of the vector space $\Sigma^n \left(G \right)$ (we used such a basis in §17 for the case $n = 3, G = D$). In other words, for every polyhedron $M \subset R^n$ there are uniquely defined real numbers $\mu_\alpha \left(M \right)$ (of which only a finite number are nonzero) satisfying the condition

$$M - \sum_\alpha \mu_\alpha \left(M \right) N_\alpha \in P_0^n \left(G \right).$$

The functions μ_α can be uniquely extended onto the space P^n:

$$\mu_\alpha \left(\lambda_1 M_1 + \ldots + \lambda_k M_k \right) = \lambda_1 \mu_\alpha \left(M_1 \right) + \ldots + \lambda_k \mu_\alpha \left(M_k \right).$$

Since these functions vanish on the subspace $P_0^n \left(G \right)$, they can be regarded as defined on the factor space $\Sigma^n \left(G \right) = P^n/P_0^n \left(G \right)$. Every function $\mu_\alpha \left(M \right)$ is an additive G-invariant and has the homogeneity property (i.e., $\mu_\alpha \left(\lambda M \right) = \lambda \mu_\alpha \left(M \right)$). In other words, μ_α is a *linear functional* on the space $\Sigma^n \left(G \right)$. It is clear that if $\mu_\alpha \left(M' \right) =$

$\mu_\alpha (M'')$ for arbitrary α, i.e., if $\mu_\alpha (M' - M'') = 0$, then $M' - M'' \in P_0^n (G)$ and hence the polyhedra M' and M'' are G-equidecomposable modulo Z_2. Thus the following theorem holds [24] : *A necessary and sufficient condition for two polyhedra M', $M'' \subset R^n$ to be G-equidecomposable modulo Z_2 (where $G \supset T$) is that $\mu (M') = \mu (M'')$ for every homogeneous additive G-invariant μ.*

The indeterminancy involved in considering "all" homogeneous additive G-invariants can be removed in the case $G = T$. In fact, reference [24] contains an explicit list of homogeneous additive T-invariants. This gives a necessary and sufficient condition, in practical form, for polyhedra to be T-equidecomposable modulo Z_2. However, it is only for $n \leqslant 3$ that the necessity and sufficiency of these conditions for T-equidecomposability (without the stipulation "modulo Z_2") have been successfully established.

Finally we observe that for an arbitrary homogeneous additive T-invariant μ and arbitrary *convex* polyhedra P and Q we have

$$\mu (P \times Q) = \mu (P) + \mu (Q),$$

so that the polyhedra $P \times Q$ and $P + Q$ are T-equidecomposable modulo Z_2.

CONCLUSION

In the table given below we summarize the basic results connected with Hilbert's third problem. The empty boxes in the table give an idea of the problems that remain unsolved. The most important of these are, without doubt, the problem of finding a necessary and sufficient condition for D-equidecomposability of polyhedra in R^n if $n \leqslant 5$ and the problem of finding a necessary and sufficient condition for T-equidecomposability of polyhedra in R^n if $n \geqslant 4$).

There are also a number of unsolved problems of lesser significance (for example, finding criteria for S-equidecomposability in space, the problem of the minimality of the group D_0 for n-dimensional spaces of $n > 3$, "minimality" questions for the decompositions used to accomplish equidecomposability, etc.) A number of results presented in this book are not listed in the table, for example, results pertaining to non-Archimedean and non-Euclidean geometries (see §8), equidecomposability with respect to the group of similarity transformations (the end of §16), tetrahedra equidecomposable with respect to a cube, and others. There are certain unsolved problems in connection with these results too (for example, the problem of equidecomposability of n-dimensional polyhedra with respect to the group of similarity transformations if $n > 3$, the problem of the existence of *uncountable* families of tetrahedra, other than Hill's tetrahedra, equidecomposable with a cube, etc.)

Nevertheless, despite the presence of empty places in the table (and other unsolved problems as well), it can now be said that *the theory of equidecomposability (and equicomplementability) of polyhedra is solved in all its main aspects and that the resulting beautiful edifice is a worthy monument to the memory of David Hilbert.*

	Equality of volume (or area) and equidecomposability	Equidecomposability and equicomplementability	D-equidecomposability and D_0-equidecomposability	Conditions for D-equidecomposability	Conditions for T-equidecomposability	Minimal group G for which G-equidecomposability coincides with D-equidecomposability
R^2	Equivalent, §7	Equivalent, §7	Equivalent, §9	Necessary and sufficient, §7	Necessary and sufficient, §10	§11
R^3	Nonequivalent, §13	Equivalent, §16	Equivalent, §12	Necessary and sufficient, §§14 and 17	Necessary and sufficient, §19	§21
R^4	Nonequivalent, §20	Equivalent, §16	Equivalent, §12	Necessary and sufficient, §20	Necessary §22	
R^n $(n \geqslant 5)$	Nonequivalent, §20	Equivalent, §16	Equivalent, §12	Necessary §20	Necessary §22	

ON THE CONCEPT OF LENGTH

In Chapter I we examined the key issues involved in defining area and volume. We now turn our attention to the concept of *length*.

The length of a line segment is defined by using the familiar *process of measurement,* which can be defined, for example, as follows. We fix a segment I_0, which is regarded as the unit of measurement. Then on a line R we fix an integer-valued scale; i.e., the set of points obtained by starting from some point $a \in R$ and consecutively laying off segments congruent to I_0 in both directions. We agree to call this scale the *zeroth scale*. The points of the zeroth scale divide R into a countable number of segments congruent to I_0, with no common interior points, called the *segments of the zeroth scale*. Dividing each segment of the zeroth scale into 10^k congruent parts, we obtain a decomposition of R into a countable number of congruent segments, with no common interior points, called the *segments of the kth scale* ($k = 1, 2, \ldots$).

Now, given an arbitrary segment $Q \subset R$, let a_k be the number of segments of the kth scale entirely contained in Q, and let b_k be the number of segments of the kth scale having a nonempty intersection with Q. Then, as is easily seen,

$$a_0 \leqslant \frac{a_1}{10} \leqslant \cdots \leqslant \frac{a_k}{10^k} \leqslant \cdots \leqslant \frac{b_k}{10^k} \leqslant \cdots \leqslant \frac{b_1}{10} \leqslant b_0.$$

It follows from these inequalities that the limits

$$\lim_{k \to \infty} a_k/10^k, \ \lim_{k \to \infty} b_k/10^k \tag{1}$$

exist. Moreover, the limits coincide because of the easily verified formula $b_k = a_k + 2$. This common value of the limits (1) is called the *length* of the segment Q, and is denoted by $l\,(Q)$.

It is not hard to prove (cf. §2) that the length of a segment (as introduced constructively with the help of the measurement process just described) has the following properties, analogous to those considered in Chapter 1 in connection with the area and volume:

(α) *The function l is nonnegative, i.e., the length $l\,(Q)$ of any segment Q is a nonnegative number.*

(β) *The function l is additive, i.e., if a point $c \in Q$ divides the segment Q into subsegments Q_1 and Q_2, then $l\,(Q) = l\,(Q_1) + l\,(Q_2)$.*

(γ) *The function l is invariant under translations, i.e., congruent segments have the same length.*

(δ) *The function l is normalized, i.e., the unit of measurement I_0 is of length 1.*

Just as the area (or the volume) of a figure does not depend on the choice of the zeroth mosaic, so the length of a segment is invariant under displacement of the zeroth scale (or under replacement of the unit of measurement I_0 by a congruent segment I_0'). The proof of this fact is much *simpler* than in the case of area or volume. The point is that any two congruent segments on the line can be obtained from each other with the help of a *translation*. Therefore, instead of axioms (γ) and (γ^*), which we had in the case of area (or volume), here a single axiom is sufficient. In other words, if a function (defined on the set of all segments of the line R) is invariant under translations, then it is invariant under arbitrary motions of the line R (i.e., invariant under reflections as well).

The indicated properties uniquely determine the length, i.e., *there exists one and only one function l defined on the set of all sequences which satisfies the conditions* (α), (β), (γ), (δ). Therefore we can replace the constructive definition of the length of a segment by an axiomatic definition: *By the length is meant a real function defined on the set of all sequences and satisfying the conditions* (α), (β), (γ), (δ).

The question of the independence of axioms (α), (β), (γ), (δ) is treated in the same way as in the case of area (§4) or volume. But here there is one subtlety. As we recall, in §10 (see the footnote on p. 81) the fact that axiom (α) is independent of (β), (γ), and (δ) was established (in the case of area) *without using* the axiom of choice. In other words, we gave a constructive definition of an additive and normalized function s_α, which is invariant under translations and does not satisfy axiom (α). Since in the case of length it is just invariance under *translations* that is at issue, one might well get the impression that the independence of axiom (α) in the case of the length of a segment can be established without using the axiom of choice. In fact, this is not the case, since the considerations given in §10 require at least *two* dimensions and do not carry over to the case of the *line R*.

We now examine this matter in more detail, and show that in the case of the length of a segment, the independence of axiom (α) can be proved *only* by using the axiom of choice. In fact, using the axiom of choice, we can construct a Cauchy–Hamel function $f(x)$ satisfying the condition $f(1) = 1$, $f(\sqrt{5}) = -1$ (cf. p. 30), and then $l_\alpha(Q) = f(l(Q))$ will be a real function on the set of all segments which satisfies axioms (β), (γ) and (δ), but does not satisfy axiom (α).

Conversely, suppose that there exists a function $l_\alpha(Q)$, defined on the set of all segments of the line R, which satisfies axioms (β), (γ) and (δ), but does not satisfy axiom (α). We introduce a coordinate x on the line R, i.e., we fix a point $o \in R$ and a direction ("to the right," "to the left") on R, and then define the coordinate x of an arbitrary point $a \in R$ by the following rule: $|x| = l([o, a])$ and $x > 0$ if and only if a lies to the right of o. Next we define a function $f(x)$ by setting

$$f(x) = l_\alpha([o, a_x]) \quad \text{if} \quad x > 0, \tag{2}$$

$$f(-x) = -f(x), \tag{3}$$

where a_x is the point of the line R which has the coordinate x. It is easy to see that the function $f(x)$ is additive. In fact, if $x > 0$, $y > 0$, we have

$$f(x + y) = l_\alpha([o, a_{x+y}]) = l_\alpha([o, a_x]) + l_\alpha([a_x, a_{x+y}]) =$$
$$= l_\alpha([o, \quad a_x]) + l_\alpha([o, \quad a_y]) = f(x) + f(y)$$

(here we use axioms (β) and (γ) for the function l_α, and also the congruence of the segments $[a_x, a_{x+y}]$ and $[o, a_y]$). Thus the formula

$$f(x + y) = f(x) + f(y)$$

holds for $x > 0$, $y > 0$, and by (3), this formula holds for arbitrary real x and y. In other words, the function $f(x)$ is *additive*. Since moreover l_α does not satisfy axiom (α), i.e., there exists a segment Q for which $l_\alpha(Q) < 0$, then, by (2), there exists a positive number x_0 for which $f(x_0) < 0$. Finally, $f(1) = 1$, since l_α satisfies axiom (δ). Thus we have established the existence of an additive function which is not linear, and this, as mentioned on p. 32, means that the axiom of choice holds. Thus, in the one-dimensional case (i.e., in the case of the length of a segment), whether or not (α) is independent of axioms (β), (γ) and (δ) depends on our choice of set-theoretic axioms (i.e., depends on whether the axiom of choice is accepted or rejected).

In conclusion, we observe that the entire book has dealt exclusively with the problem of *n-dimensional* volume of sets in *n-dimensional* space R^n (the book proper is devoted to the case $n \geqslant 2$, and this appendix to the case $n = 1$). The problem of measuring k-dimensional volume of sets in R^n, where $k < n$, is much more complicated. In particular, it is no longer possible to get along with just the axioms (α), (β), (γ^*), (δ) in defining k-dimensional volume in R^n if $k < n$. In even the simplest case $k = 1$, $n = 2$, i.e., in defining the *length* of a curve in the plane R^2, we need an additional axiom, namely the axiom of *lower semicontinuity*. Confining ourselves to the definition of length for *simple arcs*, we can formulate this axiom as follows:

(ε) *Given a simple arc L and any number* $\varepsilon > 0$, *there exists a number* $\delta > 0$ *such that* $l\,(L') > l\,(L) - \varepsilon$ *for every simple arc L' whose δ-neighborhood contains L.*

Axioms (α), (β), (γ^*), (δ), (ε) now allow us to give a constructive definition of length for curves (simple arcs) in the plane. In fact, let us agree to call a simple arc L *rectifiable* if the lengths of the polygonal curves inscribed in L are bounded from above. Then it turns out that *there exists one and only one function l defined on the set of all rectifiable simple arcs which satisfies the axioms* (α), (β), (γ^*), (δ), (ε). Here each of the axioms (β), (γ^*), (δ), (ε) (and also axiom (α) in the case where the axiom of choice is accepted) is independent of the others. In particular, this is true of axiom (ε), i.e., there exists a function l_ε, defined on the set of all rectifiable simple arcs which satisfies axioms (α), (β), (γ^*), and (δ), but does not satisfy (ε). In other words, it is impossible to construct a theory of length (for curves in the plane) by using only axioms (α), (β), (γ^*) and (δ) (without axiom (ε)). The reader can find the details in the author's article entitled "Arc length and surface area" (Encyclopedia of Elementary Mathematics, Vol. V, Moscow, 1966, pp. 88–141; in Russian).

BIBLIOGRAPHY

1. Aleksandrov, A. D., *A theorem on convex polyhedra,* Trudy Fiz.-Matem. Inst. Im. Steklova, 1933, vol. 4, p. 87. (In Russian)
2. Aleksandrov, P. S., editor, *Die Hilbertschen Probleme,* Akademische Verlagsgesellschaft, Leipzig, 1971.
3. Baumgartner, L., *Zerlegung des vierdimensionalen Raumes in kongruente Fünfzelle,* Math.–Phys. Semesterber. **15** (1968) pp. 76–86.
4. Baumgartner, L., *Zerlegung des n-dimensionalen Raumes in kongruente Simplexe,* Math. Nachr. **48** (1971), pp. 213–224.
5. Boltianskii*, V. G., *Zerlegungsgleichheit ebener Polygone,* Bul. Inst. Politehnic, Iaşi, **4**(8), 1958, pp. 33–38.
6. Boltianskii*, V. G., *Equivalent and Equidecomposable Figures* (translated by A. K. Henn and C. E. Watts), D. C. Heath and Co., Boston (1963).
7. Boltianskii*, V. G., *Equidecomposability of polygons and polyhedra,* Encyclopedia of Elementary Mathematics, Vol. V, Moscow, 1966, pp. 142–180. (In Russian)
8. Boltianskii*, V. G., *Decomposition equivalence of polyhedra and groups of motions,* Sov. Math. Doklady, **17** (1976), pp. 1628–1631.
9. Bricard, R., *Sur une question de géométrie relative aux polyèdres,* Nouv. Ann. Math. **15** (1896), pp. 331–334.
10. Danzer, L. W., *Zerlegbarkeit endlichdimensional Räume kongruente Simplices,* Math.–Phys. Semesterber. **15** (1968), p. 87.
11. Davis, H. L., *Packings of spherical triangles and tetrahedra,* Proc. Colloquium on Convexity, Copenhagen, 1965, pp. 42–51.

*Depending on the transliteration used, the name may also be spelled Boltyanski, Boltyanskiy, or Boltyansky.

12. Dehn, M., *Ueber raumgleiche Polyeder*, Nachr. Akad. Wiss. Göttingen Math.-Phys. Kl., 1900, pp. 345–354; *Ueber den Rauminhalt*, Math. Ann. **55** (1902), pp. 465–478.

13. Dubnov, Ya. S., *The Measurement of Line Segments*, Moscow, 1962. (in Russian)

14. Goldberg, M., *Tetrahedra equivalent to cubes by dissection*, Elem. Math. **13** (1958), pp. 107–109.

15. Goldberg, M., *Two more tetrahedra equivalent to cubes by dissection*, Elem. Math. **24** (1969), pp. 130–132; correction, ibid. **25** (1970), p. 48.

16. Goldberg, M., *The space-filling pentahedra*, J. Combinatorial Theory **13**, Ser. A (1972), pp. 437–443.

17. Goldberg, M., *New rectifiable tetrahedra*, Elem. Math. **29** (1974), pp. 85–89.

18. Goldberg, M., *Three infinite families of tetrahedral space-fillers*, J. Combinatorial Theory **16**, Ser. A (1974), pp. 348–354.

19. Goldberg, M., *The space-filling pentahedra. II*, J. Combinatorial Theory **17**, Ser. A (1974), pp. 375–378.

20. Hadwiger, H., *Zerlegungsgleichheit und additive Polyederfunktionale*, Arch. Math. **1** (1948-1949), pp. 468–472.

21. Hadwiger, H., *Zum Problem der Zerlegungsgleichheit der Polyeder*, Arch. Math. **2** (1949-1950), pp. 441–444.

22. Hadwiger, H., *Ergänzungsgleichheit k-dimensionaler Polyeder*, Math. Zeits. **55** (1952), pp. 292–298.

23. Hadwiger, H., *Mittelpunktspolyeder und translative Zerlegungsgleichheit*, Math. Nachr. **8** (1952), pp. 53–58.

24. Hadwiger, H., *Lineare additive Polyederfunctionale und Zerlegungsgleichheit*, Math. Zeits. **58** (1953), pp. 4–14.

25. Hadwiger, H., *Über Gitter und Polyeder*, Monatsh. Math. **57** (1953), pp. 246–254.

26. Hadwiger, H., *Zur Zerlegungstheorie euklidischer Polyeder*, Annali di Matematica, **36** (1954), pp. 315–334.

27. Hadwiger, H., *Zum Problem der Zerlegungsgleichheit k-dimensionaler Polyeder*, Math. Ann. **127** (1954), pp. 170–174.

28. Hadwiger, H., *Vorlesungen über Inhalt, Oberfläche und Isoperimetrie*, Springer-Verlag, Berlin, 1957.

29. Hadwiger, H., *Translative Zerlegungsgleichheit der Polyeder des gewöhnlichen Raumes*, J. Reine Angew. Math. **233** (1968), pp. 200–212.

30. Hadwiger, H. and Glur, P., *Zerlegungsgleichheit ebener Polygone*, Elem. Math. **6** (1951), pp. 97–106.

31. Hamel, G., *Eine Basis aller Zahlen und die unstetigen Lösungen der Funktionalgleichung:* $f(x + y) = f(x) + f(y)$, Math. Ann., **60** (1905), pp. 459–462.

32. Hilbert, D., *Foundations of Geometry*, translated by E. J. Townsend, Open Court Pub. Co., Chicago, 1902.

33. Hilbert, D., *Mathematical Problems*, lecture delivered before the International Congress of Mathematicians in Paris in 1900, translated by M. W. Newson, Bull. Amer. Math. Soc. **8** (1902), pp. 437–479.

34. Hill, M. J. M., *Determination of the volumes of certain species of tetrahedra without employment of the method of limits*, Proc. London Math. Soc. **27** (1896), pp. 39–53.

35. Jessen, B., *Orthogonal icosahedra*, Nordisk Matem. Tidskrift **15** (1967), pp. 90–96.

36. Jessen, B., *The algebra of polyhedra and the Dehn–Sydler theorem*, Math. Scand. **22** (1968), pp. 241–256.

37. Jessen, B., *Zur Algebra der Polytope*, Nachr. Akad. Wiss. Göttingen Math.-Phys. Kl., 1972, pp. 47–53.

38. Jessen, B., Karpf, J. and A. Thorup, *Some functional equations in groups and rings*, Math. Scand. **22** (1968), pp. 257–265.

39. Kagan, V. F., *Über die Transformation der Polyeder*, Math. Ann., **57** (1903), pp. 421–424; *On the Transformation of Polyhedra*, second edition, Moscow, 1933. (In Russian)

40. Kiselev, A. P. *Geometry, Part II*, Moscow, 1974. (In Russian)

41. Klein, F., *Vergleichende Betrachtungen über neuere geometrische Forschungen*, Math. Ann. **43** (1893), pp. 63–100.

42. Lebesgue, H., *Sur l'équivalence des polyèdres, en particulier des polyèdres réguliers, et sur la dissection des polyèdres réguliers en polyèdres réguliers*, Ann. Soc. Polon. Math. **17** (1938), pp. 193–226; **18** (1945), pp. 1–3.

43. Lenhard, H. -C., *Über fünf neue Tetraeder, die einem Würfel äquivalent sind*, Elem. Math. **17** (1962), pp. 108–109.

44. Nicoletti, O., *Sulla equivalenza dei poliedri*, Rend. Circ. Mat. Palermo **37** (1914), pp. 47–75; **40** (1915), pp. 194–210.

45. Perepelkin, D. I., *A Course of Elementary Geometry, Part I*, Moscow, 1948. (In Russian)

46. Rokhlin, V. A., *Area and volume*, Encyclopedia of Elementary Mathematics, Vol. V, Moscow, 1966, pp. 5–87. (In Russian)

47. Solovay, R., *A model of set-theory in which every set of reals is Lebesgue measurable*, Ann. Math. **92** (1970), pp. 1–56.

48. Sommerville, D. M. Y., *Space-filling tetrahedra in Euclidean space,* Proc. Edinburgh Math. Soc. **41** (1923), 'pp. 49–57.
49. Sommerville, D. M. Y., *Division of space by congruent triangles and tetrahedra,* Proc. Royal Soc. Edinburgh, **43** (1923), pp. 85–116.
50. Sydler, J. -P., *Sur la décomposition des polyèdres,* Comment. Math. Helvet. **16** (1943-1944), pp. 266–273.
51. Sydler, J. -P., *Quelques propriétés de la configuration complémentaire de Desargues,* Elem. Math. **10** (1955), pp. 32–37.
52. Sydler, J. -P., *Sur les tétraèdres équivalents à un cube,* Elem. Math. **11** (1956), pp. 78–81.
53. Sydler, J. -P., *Sur quelques polyèdres équivalents obtenus par un procédé en chaînes,* Elem. Math. **14** (1959), pp. 100–109.
54. Sydler, J. -P., *Conditions nécessaires et suffisantes pour l'équivalence des polyèdres de l'espace euclidien à trois dimensions,* Comment. Helv. **40** (1965), pp. 43–80.
55. Uspenski, Ya., *Introduction to the Non-Euclidean Geometry of Lobachevski–Bolyai,* Petrograd, 1922. (In Russian)
56. Yaglom, I. M., *Geometric Transformations* (translated by A. Shields), Random House, New York, 1962.
57. Jech, T. J., *Lectures in Set Theory with Particular Emphasis on the Method of Forcing,* Lecture Notes in Mathematics 217, Springer–Verlag, New York, 1971.
58. Zorn, M., *A remark on method in transfinite algebra,* Bull. Amer. Math. Soc. **41** (1935), pp. 667–670.
59. Zylev, V. B., *Equicomposability of equicomplementable polyhedra,* Sov. Math. Doklady, **161** (1965), pp. 453–455.
60. Zylev, V. B., *G-composedness and G-complementability,* Sov. Math. Doklady, **179** (1968), pp. 403–404.

SUBJECT INDEX

WITHDRAWN